Out of Character

Out of CHARACTER

Surprising truths about the **LIAR, CHEAT, SINNER (AND SAINT)** *lurking in all of us*

DAVID DESTENO *and*
PIERCARLO VALDESOLO

THREE RIVERS PRESS • NEW YORK

Published in the United States by Three Rivers Press,
an imprint of the Crown Publishing Group,
a division of Random House, Inc., New York.
www.crownpublishing.com

Three Rivers Press and the Tugboat design are registered
trademarks of Random House, Inc.

Originally published in hardcover in the United States by
Crown Archetype, an imprint of the Crown Publishing Group,
a division of Random House, Inc., New York, in 2011.

Library of Congress Cataloging-in-Publication Data
DeSteno, David.
Out of character : surprising truths about the liar, cheat,
sinner (and saint) lurking in all of us / David DeSteno &
Piercarlo Valdesolo.—1st ed.
p. cm.
Includes bibliographical references and index.
1. Character. 2. Conduct of life—Psychological aspects.
I. Valdesolo, Piercarlo. II. Title.
BF818.D43 2011
155.2—dc22 2010050275

ISBN: 978-0-307-71776-4
eISBN: 978-0-307-71777-1

Printed in the United States of America

10 9 8 7 6 5 4 3 2 1

First Paperback Edition

To our wives, Amy and Liz

Contents

Out of Character

1 / Saints and Sinners

The mental battle that defines our character

Marshall Clement Sanford was an Eagle Scout—literally. The son of a respected Florida family, in his younger years he was a proud member of Boy Scout Troop 509 from Pompano Beach. But these early years weren't all campfires and fishing trips—becoming an Eagle Scout was hard work, both physically and mentally. Besides learning how to tie knots that would hold your weight, mastering the exacting science of log cabin construction, and figuring out which way was north based on the position of the sun, being an Eagle Scout also constituted, as Marshall would later say, an important and arduous developmental voyage—a voyage "of character, leadership, and persistence."[1]

For him, it was a voyage that appeared to pay off. Marshall did well for himself. After high school, he graduated from Furman College at the top of his class. He went on to complete an MBA at the University of Virginia's prestigious Darden School of Business and a summer internship at Goldman Sachs. Marshall was rising fast, widely admired for his skills, for his smarts, and for being a straight shooter. That summer of his Goldman internship was also the summer he met Jenny Sullivan at a party in the Hamptons, and when he returned to New York City that fall to take a high-profile job, he promptly asked her to marry him. Although they both had promising careers in the big city—Jenny was a vice president of a large investment firm—they soon decided to move back to South Carolina (where Marshall's family had moved his senior year of high school), where they could live a more traditional life.

Once settled back down South, Marshall headed up a real estate company and Jenny raised their four boys in what everyone agreed was the picture of family harmony. Although her husband could be quirky and often a bit stoic, she admired him for his honesty and integrity. As Jenny would later say, "He cherished Galatians 5:22: 'The fruit of the Spirit is love, joy, peace, patience, kindness, goodness, faithfulness, gentleness, and self-control,'" and he lived accordingly.[2] But Marshall was also ambitious and passionate about serving his community. So the beloved native son, the very embodiment of strong moral character and good old-fashioned American values, decided it was time to run for public office.

Okay, we know what you're thinking: good character and politics aren't usually two things that go together. But Marshall was not your average politician. He wasn't in it for the prestige, the perks, or the power. He liked to describe himself as a "citizen legislator" who was in it to do the right thing by South Carolinians—to be a champion of the people. A political neophyte with a fresh face and a boatload of enthusiasm, his straightforward and earnest demeanor catapulted him to victory in his first run for Congress in 1994, where he served three terms. Those terms weren't tarnished by scandal or ego or disgrace, as they are for so many in his line of work; instead, he was widely seen as a staunch advocate and strong voice on issues of both social and fiscal responsibility. But he didn't just vote his values, he lived them. He not only fought wasteful spending during the day but was just as judicious with his own money—and the taxpayers'—at night. With little interest in the material excesses or extracurricular dalliances of the Washington, D.C., party scene, he spent his nights in the capital on the futon in his office, accommodations he preferred to renting an apartment on the government's dime. Conservative both in lifestyle and in politics, his straight-arrow persona made him a conservative favorite back home in the red state of South Carolina, and as a result, by 2003 he, Jenny, and the boys found themselves moving into the governor's mansion.

It was a welcome change for the family, as living apart had been difficult, with the frequent separations limiting the couple's time for deep conversation and sharing the ups and downs of daily life. But now everything was again falling into place. "Though we were both incredibly busy, we'd been living under the same roof at last, and with that proximity," Jenny said, "I'd fallen in love with him all over again."[3] And so, it seems, had his constituents. From the very outset of his term, Marshall was trumpeted both in his home state and in Washington as a new kind of politician—a man of virtue. Even if you didn't agree with his policies, there seemed to be no question that he was a good man.

Yet on June 24, 2009, Marshall "Mark" Sanford's life changed forever. Upon arriving back in the United States from a trip to Buenos Aires, he was met by a reporter who, like many South Carolinians, had spent the past week wondering about Sanford's whereabouts. The governor had gone AWOL, offering his staff, his family, and his constituents only the flimsy lie that he was hiking the Appalachian Trail. But as we now know, he was actually in Argentina with his mistress—or his "soul mate," as he would later call her. It turned out that the seemingly levelheaded and loyal governor had been penning erotic love letters to Maria Belén Chapur for months. Evidently he had just returned to the States with more material about which to write.

Mired in a tug-of-war between his firmly held convictions about what was "right" and his desire for the woman he now claimed was his once-in-a-lifetime love, Sanford, in a tear-filled press conference later that day, begged forgiveness for his moral transgression, admitting that he had crossed the "sex line" and apologizing for the pain he had caused. But it was too late. On that day Mark Sanford's image suddenly changed forever. He was no longer a paragon of virtue, and his political ambitions, along with his character, were consigned to the junk heap.

The good and bad in all of us

Cases such as Sanford's—and the many others like it that regularly grace the headlines—fascinate us. The idea that a person seemingly living a life of propriety could commit such shameful acts, along with the suggestion that we could be so easily fooled by the pretense of goodness, shatters our confidence in our ability to judge others—or even ourselves—accurately. Whether the transgressor is a politician touting family values while carrying on an affair with an international mistress, the next-door neighbor who "seemed just like everybody else" until he committed an act of terrorism as a member of a radicalized political group, or the admired and upstanding hedge fund manager who turned out to be the perpetrator of a multibillion-dollar Ponzi scheme, when people act in a way that violates our expectations and beliefs about their character, we—both as individuals and as a society—are often shaken to our very core. To compensate for our errors in judgment, we convince ourselves these people must have been wolves in sheep's clothing—inherently nasty individuals who may have managed to hide in plain sight for a time, but whose true colors have ultimately been revealed. Hindsight, after all, is 20/20. We tell ourselves that Sanford's fall from grace must have been long in the coming. He must have had some flaw in his character that lurked there those many years, hidden behind that Eagle Scout badge, something that Jenny (and the rest of us) just couldn't see. If we had just looked closely enough, maybe there would have been clues, windows that would have let us discern who Sanford really was as opposed to who he presented himself to be. How else could a man who once seemed such an exemplar of good character have turned out to be a lying, cheating philanderer? How else could we all have been duped?

These are good questions. But the answer, we'll argue, is not that we missed some telltale signs or that we are gullible fools. No, it's not that we misjudged his character; it's that our understanding of the *concept* of "character"—what it actually is and how it works—is fundamentally wrong.

Character—what *Webster's* defines as "the complex of mental and ethical traits often individualizing a person"—has long been almost universally agreed to be a stable fixture. People believe that it is formed at an early age through learning and experience, and that it becomes internalized and solidified into a deep-seated disposition that guides their actions over the course of their lives. In fact, the word *character* itself comes from an ancient Greek term referring to the marks impressed indelibly upon coins to tell them apart. And since that time, the term has been used to describe the supposed indelible marks pressed upon humans' minds and souls that "reveal" their true nature. Character is the currency we employ to make judgments about people—to determine who is good and who is flawed, who is worthy and who is not, who is saved and who is damned. Character, quite simply, is who we are, like it or not. Everyone believes this to be a fact; even *The Complete Idiot's Guide to Understanding Ethics* says that character traits are fixed, deeply ingrained features of personality.

But if this view is correct, some things just don't add up. If character is stable, how could Mark Sanford and others like him fool so many people for so long? How could they have concealed their moral shortcomings from their families, friends, colleagues, and communities year after year? It's hard to imagine that most people are capable of such an elaborate ruse. As Tom Davis, one of Sanford's closest friends for thirty years, put it: "I've known Mark, and the opinion I've formed of him, I never would have expected something like this. This is not in character for Mark Sanford."[4] Virginia Lane, one of Jenny's close friends, echoed the view: "Mark's the last person on the planet we thought this would happen to."[5] And Jenny herself was the most shocked of all: "I always believed that Mark and I had no secrets. After all of these years in the public eye, our lives were open books to one another, let alone the public."[6] "It never occurred to me that he would do something like that," she said upon reflection. "The person I married was centered on a core of morals."[7]

But in a way, our responses to situations like these aren't entirely logical or fair. Should a single moral failing erase a lifetime of good behavior? Why does a single transgression seem to give us license to brand someone with the indelible mark of a marred character? One explanation is that because these single events are so shocking and so memorable (not to mention so beaten to death by the media), they eclipse all else. But if you buy that view, then why isn't the reverse true? Why doesn't a single good deed, even a memorable one, ever seem to be seen as a mark capable of defining a person's true colors? Ever heard of Farron Hall, the homeless alcoholic who lived under a bridge in Winnipeg, and who in May 2009 risked his own life by jumping into the Red River in a heroic attempt to save a drowning teen? Probably not. That's because despite risking his life to save a total stranger, he was never hailed as a role model, never awarded a medal of honor or invited on the talk show circuit to discuss his moral bona fides. Instead, he was patted on the back by local officials and quickly forgotten. In society's eyes, this one good act wasn't nearly enough to redeem Hall from a lifetime of "degenerate behavior."

It seems that wolves may masquerade as sheep, but sheep just don't masquerade as wolves. We rarely view one good act as proof someone had good character all along, yet most of us are ready and willing to do the reverse. Those marked as "bad" can do something nice now and again and our opinion of them doesn't change, but all it takes for a person of seeming high virtue is one slip for us to claim that his or her character is inherently flawed.

This double standard may not be fair, but it's also not particularly surprising. As work by the psychologist Paul Rozin has shown, humans possess a fundamental tendency to accentuate the negative.[8] Drop a fly into a bowl of delicious soup and the soup suddenly becomes inedible. Yet placing a drop of delicious soup in a bowl of dead flies hardly makes for a tasty treat. This may be an extreme example, but the point is that, rational or not, for the mind any sign of contamination—physical or moral—is hard to ignore. History has borne this tendency out over and over. In one particularly egregious example, during the nineteenth and early twentieth centuries it was accepted in many southern states that a single drop of "black blood" in one's ancestry rendered one legally black, therefore tainting and making one ineligible for all the civil rights that applied at the time, whereas the reverse didn't apply. In short, the things we deem "bad" consistently seem to hold more weight than those we deem "good."

This very fact provides a bit of a problem for the commonly held view of character as a stable phenomenon. Think of it this way: if you believe that character is fixed, you have to accept that an instance of behaving "out of character" is one of two things: (1) an aberrant event (like Hall's heroic act) or (2) a window into the person's "true" and yet hidden nature (like Sanford's indiscretion). But in reality, which one we choose seems to depend on whether the person in question was a "saint" or "sinner" to begin with.

An even bigger problem for the fixed view of character is that acting "out of character" isn't a freak occurrence or something restricted to the famous few. As we'll see throughout this book, it's actually much more commonplace than most people think. There lurks in every one of us the potential to lie, cheat, steal, and sin, no matter how good a person we believe ourselves to be. Combine these two problems, and the view of character as a stable fixture begins to crumble.

This is not to say that character doesn't exist or that our behavior is completely unpredictable. A random system like that wouldn't make any sense either. If the mind worked that way, our social world would be chaos—our actions at any moment in time would be reduced to a simple roll of the dice. No, character exists. It just doesn't work the way most people think. In the chapters that follow, we'll show you that hypocrisy and morality, love and lust, cruelty and compassion, honesty and deceit, modesty and hubris, bigotry and tolerance—in short, vice and virtue—can coexist in each of us, and that the behavior or decision that emerges in any given moment or situation isn't necessarily the one we intend. Yet the decisions we make and the actions we take aren't haphazard; they're the product of dueling forces in our minds. As with most duels, however, there are a set of rules that guide the moves of the combatants. But to fully understand these rules, and to learn what you can do to guide the outcome of the battle, you first have to be willing to give up everything you thought you knew about character—what it is, how it is formed, and how it works.

A new look for an old problem

When most people think of the dueling forces that shape who we are, they picture that familiar image (often in cartoon form) of the angel on one shoulder and the devil on the other. You know, that image of two little yous—one dressed in a long white robe and a halo and the other in a body-hugging red number with horns and a pitchfork—sitting on your shoulders and whispering intently into your ears. These two little guys (or girls) have long been used to represent conflicting forces—one urging good and the other bad—that occupy our subconscious and try to influence our every decision.

Devil: "Eat all the cookies, then blame it on your brother."
Angel: "No, don't eat all the cookies, and make sure you thank Grandma for baking them."

Our character, then, is thought to have much to do with which one of these voices we tend to favor over the other—as a general rule, are we likely to eat all the cookies or thank Grandma? What's more, the voice we favor is assumed to be determined at a young age, influenced by (deliberately or not) our parents, teachers, peers, and the like. So by the time the dilemmas we face get a little more complicated—from cookie jars to dalliances with mistresses and the whole host of other moral lapses that get adults into trouble—it was believed the battle between the little shoulder sitters had pretty much reached détente; the voice we were going to listen to had already become a foregone conclusion. "Character," as Plutarch noted centuries ago, "is a habit long continued."

Now, we're simplifying a bit, as there are ongoing debates about the roles of temperament, culture, religion, upbringing, and other social and environmental factors in shaping character. But wherever you come out on these issues, most theories reach the same conclusion: good character is developed by deciding, early in life, to favor the "good" voice over the "bad" one, or in other words, by consciously deciding to tamp down those craven impulses that want all the cookies, the money, or the sex—and want them now. Only by exerting willpower, it was believed, could one cultivate a noble spirit capable of ignoring sinful temptations and evil urgings. And that once that virtuous voice won out, it would become etched in your psyche and your character would be set forever.

This view seems logical. It feels right. It fits with all our other preconceived notions about ourselves. There's just one problem: it isn't borne out by the data. And it doesn't explain why humans are prone to behaving "out of character" as frequently and seemingly unpredictably as they do, sometimes in ways that surprise even themselves. Simply put, the view is fundamentally incorrect.

In the chapters that follow, we'll explain, using a wealth of rigorous psychological research, exactly why this long-held theory is flawed, and in so doing, we'll argue for a new, more enlightened view of character that takes into account how people actually act. We'll show why the mind values flexibility and why, whether we like it or not, exerting willpower doesn't inherently make us an angel, any more than indulging our urges necessarily makes us a devil. Life isn't that simple. Navigating the social world successfully involves being able to adapt our behavior to the challenges and opportunities that individual situations present. A character that is fixed or etched—and by that we mean a mind that consistently and automatically forces us to listen to one "voice" over the other—couldn't possibly steer us through the complex world of human social relations. After all, one of the greatest evolutionary advantages humans have is cognitive flexibility, the ability to fine-tune our thoughts, attitudes, and behaviors in the face of changing contexts and situations. Just as we are wired to recognize when something that may usually be bad for us (say, drinking a foul-tasting substance) can be, in a specific instance, good (drinking foul-tasting medication), so too can we recognize when a social act that is generally advantageous (for example, being generous or telling the truth), can in certain cases lead to problems (as, for example, when our generosity causes us to be taken advantage of, or when the truth unnecessarily hurts someone's feelings). When we seem to act out of character, then, it's not because we've just had a mental hiccup or we let our guard down; it's because in that moment or situation, our actions, at least to some part of our minds, seemed optimal. The problem, though, is that *seeming* optimal and *being* optimal can be quite different things.

So how are we to understand the malleability of character?

The best way is to envision character as a fluctuating state, not a permanent trait. It's not a static attribute like blue eyes or broad shoulders; it's a state that is always shifting, trying to find the right balance between competing psychological mechanisms. Picture a scale, the old-school kind with golden plates on each side. At any point in time, the scale can be balanced in infinite ways, from a level position with equal weight on both sides to heavily tipped with many more weights on one side than the other. Character is much like that scale—how a person acts at any moment is determined by how the scale is tipping, or where along the continuum it's balanced at that exact moment. Character unfolds over time, but not in a slow or linear way. The scale can shift, and shift quickly, in either direction. In fact, it's constantly oscillating to adjust to our needs, situations, and priorities. And the direction in which it shifts in any given moment is determined by the outcome of the struggle between dueling mechanisms in our mind.

Now, you might be wondering whether this metaphor isn't just like the angel-vs.-devil one, except that it's a little more fluid. Isn't it essentially saying that we're continually choosing between dueling voices in our head, one telling us to be good and the other bad? Simple answer: definitely not. The angel-vs.-devil view isn't only wrong, it's wrong in three big ways. First, the dueling voices aren't good and bad. It's much more nuanced; the little guys don't wear horns or halos. Second, it's never certain which voice to trust. In the old view, we learned early on in life that one voice would bring us more happiness, so we simply decided to always trust that voice and willfully ignore the other. Well, as we'll show throughout this book, not only is it unclear which voice has your best interests in mind, it's also unclear if you can even trust yourself, or your gut for that matter, to decide which one to heed. Both your reasoned thoughts and your intuitions will try to tip you to one side or the other, and sometimes they pull you in different directions. Third, the fight being waged within usually isn't a fair one. For several reasons we'll look at later on, the starting point for the scale, as well as the strength of the mental mechanisms pulling down on each side, can be easily manipulated by external forces, even without our realizing it. The slightest shift in a situation can pump one side with steroids, so to speak, pushing the scale in its direction. In short, character isn't decided by a simple one-off. The story of how we get to be the kind of person we are just got a lot more complicated.

Good vs. bad is so passé

Good vs. bad: it seems pretty clear what side you should be on, right? Sure, it might be fun now and again to take a walk on the wild side, but everyone knows that virtue is the way to live the best life. At least that's how the story goes. The only problem is, it's just a story. It might work well for fairy tales and fables, but under the scientific lens it just doesn't hold up. For most people, *virtue* implies attributes such as honesty, compassion, generosity, and humility, while *vice* implies the opposite. Just take a look at the seven deadly sins. There's lust, gluttony, greed, sloth, wrath, envy, and pride. The corresponding virtues are chastity, temperance, charity, diligence, patience, kindness, and humility. Always striving for the good seven and avoiding the bad ones surely will make you saintly, no question about it. But as for happy and successful (in the psychological, material, and biological senses), we're not so sure.

Take generosity and kindness. Sure, these are great ways to behave—to a point. The overly generous can give too much or to the wrong people; the unflinchingly kind can sacrifice their own well-being (or that of their family) just for the sake of being nice. Likewise, in too large doses, humility can leave you stranded on the low rung of the corporate ladder. Charity can wipe out your savings. Patience can leave you waiting in the wings indefinitely. And as for chastity and temperance . . . well, those downsides are obvious. The serious point here is that if taken to extremes, these "virtues" can be quite problematic. If giving to others means depriving yourself too much, and if caring about others' needs means always sacrificing your own, nobody wins, because a population of pure altruists simply isn't sustainable. It sounds cynical, but sooner or later the drive for self-interest will kick in one way or another, and the most virtuous will get left behind.[9]

Similarly, the so-called vices aren't always as bad as one might think. For example, pride can motivate us to develop more skills, spur us to make useful contributions in the workplace and in our communities, and mark us as potentially strong leaders. Wrath and anger can sometimes be the fuel we need to ensure fair play and to fight for things we (and others) deserve. After all, few people would blame a man (or woman) for seeking to punish those who hurt him or those he loves. In fact, not doing so would be seen as a character flaw in many cultures. Lust might be what first attracts us to a future spouse—the mother or father of our future children. In short, the feelings and behaviors that are generally considered to be "vices" don't always lead to ill.

A more accurate way to understand the complex battle underlying our social behavior is not to think of the scale of character with an angel on one side and a devil on the other, but rather to use a different metaphor—the ant and the grasshopper. If you remember Aesop's fables, you'll recall that the story describes two insects with very different predilections. The ant is always looking to the future—it would rather toil to store away food for winter than enjoy the balminess of a midsummer day at leisure. The grasshopper, on the other hand, sees no point in worrying about the future until it gets here, so it spends its time singing, playing, and enjoying itself. *Carpe diem* is its motto, at least until autumn comes.

Now, picture the scale of character flanked by these two. On one side we have the mental systems that focus on immediate rewards, or pleasure, in the short term: the grasshopper. On the other side, we have the systems that focus on long-term concerns, or what's best for the future: the ant. The important point to realize, though, is that it's not the case that the grasshopper is always a force for vice. *Both* the systems of the ant and the grasshopper are looking out for our best interests; they just do so in different time frames. Now, that said, in the fable, the moral is clear: it's better to be like the ant and always be prepared for the future. But while this may be true for ants and make for a nice story, for humans in the real world it's not so simple. You see, for ants, life is all about the long-term survival of the colony—individuals don't matter much. That's not how it is for us. Sure, it's important to look out for our long-term survival by working to be valued by our peers and acting in ways that foster social connections, but it's also important to know when great benefits can come from acting in ways that give us advantages in the here and now. To thrive, then, we need to consider the implications of our actions not just for our reputations, social standing, and ultimate well-being in the long term but also for what we can gain in the short term. Balancing these two is often a tricky business because our short-term interests frequently conflict with our long-term ones, leaving the systems of the ant and grasshopper at odds.

So if we're to use this metaphor, the shortsighted systems of the grasshopper are the ones steering you toward actions and decisions that will bring immediate pleasure and reward. This is the voice telling you to eat the cupcake, buy the new car, screw over a colleague to curry the boss's favor, or play the lottery because you just might win. We know, none of these things sounds particularly virtuous. But hold on a minute. The grasshopper is also the same voice telling you to go demand a promotion from your boss, risk your safety to protect your child, or have spontaneous sex with your spouse—all short-term urges that can contribute to physical, financial, and psychological happiness. If this sounds counterintuitive, it really isn't. Evolution has programmed us to want certain things—fatty food, sex, power—in the short term because they have the potential to increase our evolutionary fitness (i.e., the ability to thrive and thereby raise offspring to the age of sexual maturity). That's why all these things feel good to do, to have, or to consume.

But that's not the end of the story. As we hinted, humans, after all, are a social species. As a result, our evolutionary fitness also depends on having strong long-term relationships—relationships for sharing resources, for raising offspring, for defending against enemies. In fact, long-term stable relationships—or what is often termed social capital—has been shown by the psychologist John Cacioppo and others to be one of the central factors underlying human well-being.[10] This raises a problem, however. It's hard to have stable, reciprocal relationships when you're focused only on your own short-term goals. Something, then, has to counteract those short-term, self-focused impulses.

Here is where the systems of the ant come in. They recognize that reaping rewards in the future often requires making sacrifices in the moment. In other words, this is the voice telling you to repay a loan from your friend instead of using that money to buy an iPod, to spend long hours to hone a skill rather than loafing on the couch, to resist the urge to remove your wedding ring and flirt with that hot guy in the bar. Such decisions are surely less rewarding in the short run, but in the long run, they can clearly be beneficial. Your friends will trust you more and continue to share economic resources, you will gain skills that make you an attractive partner or member of society, and your romantic relationship will continue, increasing the chance of raising successful offspring.

As we've said, though, when it comes to character, nothing is black or white. Focusing too much on these long-term rewards can be problematic as well. What if your relationship is already on the rocks and keeping on that ring means you miss meeting the one person (harking back to Governor Sanford) who really is meant to be your "soul mate"? Or what if your single-minded focus on honing some skill makes you miss an experience that makes life worth living? Yes, saving money is a good thing, but never spending can make for a pretty mundane life. Likewise, working hard is admirable, but dedicating yourself solely to the office can deprive you of time with family and friends. Being completely ruled by the ant may seem virtuous at first blush, but it may not always lead to the best-lived life. As research by marketing professors Ran Kivetz and Anat Keinan shows, people can come to regret decisions that lead to an overemphasis on long-term outcomes.[11] For example, in one study, Kivetz and Keinan found that immediately after returning from winter break, college students reported regretting that they didn't study enough during the break. However, a year later, their primary regret was that they didn't enjoy themselves more when they had the chance. What's more, when Kivetz surveyed Columbia University alums returning for their fortieth reunion, he found that many, even though they had quite successful careers, wistfully reported missing out on some of the pleasures of life. If you're always saving, working, and putting off pleasure for a rainy day, you might just be old and alone by the time that day comes.

So you can see why the ant and the grasshopper are often locked in an ongoing struggle—one that usually occurs outside your awareness—to tip the scale, and thereby your decisions and actions, to their side. They both think they're right in any given situation. Yet at different times and for different reasons, each can serve you well or lead you down the road to ruin. How to decide which to listen to in any given instant? Read on.

Think or blink? Both can screw you up

Historically, "good" character has been linked with rational thinking and self-control. For the Greek Stoics, for example, virtue came from self-discipline; it sprang from the ability to resist the temptation of life's sensual pleasures. Almost two millennia later, Kant took a similar view. Virtue, for Kant, meant bringing all one's mental faculties under control and using free will as "a power to choose only that which reason, independently of inclination, recognizes to be practically necessary, that is, to be good."[12] In essence, all it took to be a good person was to figure out which course of action was best and then make yourself do it. Even in modern psychology, until quite recently anyway, this was the prevailing view: that good character came from learning early on how to silence the irrational voice steering you toward your baser impulses.

Although much research supported this view, the classic demonstration comes from the psychologist Walter Mischel.[13] This series of studies, which began in the 1960s and became well known as the "marshmallow test," looked at the relationship between self-control and social success. The experiment was as simple as it was elegant. Mischel wanted to examine the mechanisms that gave rise to willpower—the psychological processes that made some children better able to resist temptations than others. But to conduct this research, he first needed a temptation. Simple enough—what kid doesn't like marshmallows? Second, he needed a setup that would allow him to measure willpower, or how well the little ones could resist the sugary goodness. So he had a researcher place a marshmallow on a table in front of each child. The researcher then had to "step away for a few minutes," and made the child an offer: he or she could eat the one marshmallow right away, or wait a few minutes and receive *two* marshmallows. A simple choice: one now or two later. But maybe not so simple. If you ever have the opportunity to see the videos from these experiments, you'll see just how difficult it was for many of these kids to forgo the treat. Some kids, of course, didn't even try, and gobbled straightaway. But others visibly exerted tremendous psychological effort as they struggled to resist. Still others came up with rather creative strategies. A few covered their eyes with their hands (presumably assuming that what they can't see can't tempt them), and others licked—but didn't eat—the treat.

The upshot of the studies, though, was this: when Mischel followed up with the same individuals—now no longer children—more than a decade later, he found that the kids who had been able to delay the gratification of eating the first marshmallow were the ones who had the most social success later in life. They were viewed as more honest, had stronger relationships, and had achieved more academic success than their gluttonous counterparts. Good character, the researchers claimed, was linked to the early ability to resist and control impulses—to ignore the grasshopper whispering its cravings into one's ear. But that wasn't the whole story.

Economists have a fancy term for decisions like this one that hold different consequences as time unfolds: intertemporal choice. Although not always as clear-cut as with the marshmallows, these types of choices abound in our daily lives. Should I party tonight or stay home to study for the exam? Should I take the $100 I have now and splurge on something I'll enjoy or should I invest it for future gain? Should I love the one I'm with or wait to meet the one I might truly love? In all cases, we logically know that choosing the second option over the first would yield greater rewards in the long term: getting into a better college, having more money when you may need it in the future, forging a stable and loving relationship, et cetera—but people don't always make that choice. Why? Partly because the short-term rewards are often too seductive to resist. Their appeal, studies have found, lies in our innate tendency to underestimate the value of future gains relative to immediate ones—what the economists call temporal discounting. Rationally, this tendency may make little sense, but there is actually an evolutionary logic behind it. Sure, getting $200 in four weeks is usually better than $100 today (a 1,200 percent annual interest rate is hard to beat). But consider the fact that thousands of years ago, we had little way of guaranteeing that we'd see that $200 (or, more precisely, a similar type of reward) a month later. After all, there weren't any stocks, bonds, or legally binding contracts back then, nor could people then (or even now, for that matter) be sure that they'd be around thirty days later. The mind had to find ways to balance the risk of losing an immediate reward against any potential future gains, and therein lies the power of our short-term impulses.

The marshmallow studies are wonderful—clever, methodical,

and significant. And they capture the essence of the dueling forces we've been talking about. But they don't paint the whole picture. The results of the marshmallow test suggest that the way to solve the problem of intertemporal choice in the modern world is to use willpower to control our impulses and instincts.[14] And it's true that this strategy works sometimes. But it can't, and doesn't, work all the time. As we'll see throughout this book, sometimes heeding our impulses, going with our gut feelings, actually leads to *better* long-term results.

How do we account for this? Well, again, let's consider it from an evolutionary standpoint. On the evolutionary scale, self-control and discipline are relative newbies. The ability to reason about abstract concepts and weigh the different possible trade-offs of our actions stems from parts of the brain that were far less developed in our ancestors. Yet problems that require taking a long-term view did exist even if the faculties to consciously assess their consequences didn't. To survive, our progenitors regularly faced challenges requiring cooperation, fairness, reciprocity, and altruism. In some situations they needed to act selflessly in order to maintain interpersonal relationships and avoid the probable doom of social isolation. Consequently, if the only way to delay immediate gratification or short-term impulses were to bring logical reason and self-control to bear, we never would have made it out of the ancestral savannah in the first place. Which is why evolution provided the mind with competing instincts and intuitions—some focused on gain in the here and now, others urging us to delay gratification and focus on what is to come.

What we often call our gut feelings, our intuitions, are really urgings of an older mind trying to push us in a given direction. We might not even notice these feelings, as often they seem quite in accord with what we consciously believe we should do. For example, when you consciously decide that you shouldn't lie to someone, you aren't necessarily aware of that guilty feeling that was lurking underneath to push you toward honesty. However, in those instances where what we *feel* we should do and what we *think* we should do are at odds, the urgings of the older mind are hard to ignore. This is why even when you consciously decide it might be in your best interests to lie ("I believe that $100 bill you found on the floor *is* mine" or "No, I wasn't flirting"), you usually have to work hard to ignore that pang in your gut.

Religious dogmas and ethical philosophies are some of the tools the conscious mind uses—or at least thinks it does—to choose how to behave with respect to issues of character. Those of you who, like us, spent many long hours in Sunday school (or some other religious instruction) probably learned that these belief systems are usually aimed at tamping down those "irrational" and "selfish" impulses. Giving in to those urges, you probably were told, can only get you into trouble—especially when those urges are adolescent ones. But as we noted, from an evolutionary point of view, if these intuitive urges were solely selfish, they wouldn't have been doing their job. No, these older systems had to balance long-term and short-term gains just like our more recently evolved rational mind does. So while we may feel an urge to lie to protect our interests, we also feel guilt at so doing. While we feel anger or disgust toward certain people, we also feel compassion when we see others in pain. These emotional responses are the currency of the older, intuitive mind—they are the automatic engines that push us to behave one way or the other.

You can think of these feelings as the opening gambits of the ant and the grasshopper. They represent the initial position of the character scale based on an ancient and intuitive calculus. Yet immediately following this first move, the battle to tip the scale continues as we bring to bear conscious will and analysis. Conscious reasoning, however, takes a little time, so how we act in any given situation is partly determined by when in the decision process we act. The more rapidly we act—the more quickly we make a decision—the more our behavior will be influenced by the intuitive systems. The longer we take to think about it, the more conscious motives and analysis come into play, for better or for worse.

For example, if our homeless hero Farron Hall had taken even a few extra seconds to decide whether or not to jump into the Red River, he might never have risked his life to save a complete stranger. Instead, he listened to his instincts—ones born out of a venerable algorithm tipping us toward the long-term benefits of altruism—and acted before he could reflect.[15] True, poor impulse control may have created a whole host of problems in his daily life—addiction to alcohol and the like—but in that one moment, heeding his intuitions steered him toward a truly selfless act.

Before we turn to the third big problem with the common view of character, we want to emphasize one final and important point. Recently there has been an ongoing debate (fueled largely by the booming field of behavioral economics and the popularity of books on decision making) over whether it's better to trust judgments that are consciously reasoned or intuitive ones that occur in a blink. Well, when it comes to character at least, the answer is both and neither. In actuality, the question itself is misguided. You see, both the older and newer "minds" developed to serve the same goal: balancing long-term and short-term interests. Reason is a newer tool on the evolutionary continuum, but it serves the same master as instinct does. Although the ability to reason brings innumerable benefits, it's no guarantee of virtue. After all, many malevolent or dishonest acts can be justified if you're willing to engage in some "reasoning." As you'll see many times throughout this book, "irrational" or intuitive mechanisms don't always lead to the best results, but they don't always lead to misguided ones either. The same goes for so-called rational mechanisms. Remember, the struggle between the ant and the grasshopper plays out on the battlefields of both the ancient and modern minds, and neither side always holds the better answers.

It's usually not a fair fight

At this point, we've described how character may be better understood not as a set of fixed traits but rather as a temporary state—like a tug-of-war with short-term interests on one end and long-term interests on the other. We've also argued that neither intuition nor reason is always optimal—both can lead you astray. Okay, but there's one more kink in this system to consider. In the old view of character, the battle between the angel and devil was fought on a level playing field, with both having equal opportunities to present their case. If only life were that fair!

As we'll see throughout this book, these dueling psychological forces aren't always equal in strength. More often than not, one has an advantage over the other, but it's also a fragile balance of power that can shift from one minute to the next, depending on the situation or context. For example, most of us would feel a pang of guilt over cheating someone in a business deal, right? That's because our inner ant knows that in the long run, being known as a cheater likely will come back to haunt us. The scale tips toward the ant. But what if, at the moment just before you were confronted with a decision about whether to cheat, you watched a funny clip on YouTube? The good feeling the short clip produced can trick the mind by counteracting the pang of guilt that would normally accompany your deceit, and in so doing, it can aid the grasshopper in its efforts to pursue expedient cheating.[16] The scale now moves the other way. Shifts in this balance of power can come from internal changes as well—as when raging hormones make you suddenly feel powerless to resist the person you have your eye on. The point here is that the mind is subject to many sources of bias. Simple exposure to extraneous cues or information can influence our decisions without us even realizing it.

It's these kinds of cues, whether they're relevant or not to the decision at hand, that can determine which of the dueling mechanisms is more powerful at a given time. And just as the balance of power moves back and forth, so too do the behaviors that mark our "character."

Character is as character does

So what is character, then? That's a question we hope to have answered by the time you finish this book. At this point, we realize that we've asked you to take a lot on faith. We've asked you to accept that character isn't a fixed, deep-seated trait but rather a variable state; that a dishonest act doesn't make a person dishonest across the board; that the moral mind isn't subject to saintly and sinful urges; that neither reason nor intuition always provides the best answers. And finally we've asked you to accept that many of the decisions people believe reflect character are actually swayed by external forces of which they are not aware.

We don't expect you to take our word for all this on argument alone, and we wouldn't want you to. We're not philosophers—we're scientists. And as scientists, we find truth about human behavior by putting people in controlled settings where we can manipulate a number of variables and study what they actually do. Even the most logically beautiful theories or thought experiments can't hold a candle to real-world data.

We take pride in designing experiments that provide a window into the best and worst sides of human nature—situations that, even though conducted in a lab, come as close as humanly possible to replicating the ones that people confront in their daily lives. We've made it our life's work to find out, scientifically, why people choose to do what they do—and we're not above using a little trickery to get the information we're after. This isn't just to make our experiments more fun or interesting (though that's an added bonus); it's because psychologists have found that people don't act naturally in the lab if they know what behavior is being studied. So we've told some tall tales and put some unwitting folks in the middle of some elaborately staged scenarios and conflicts, but it's all in the name of science. Procedures are always cleared by review boards, and participants are always made aware of the deception at the experiment's end. No one has gotten too mad at us yet. More often than not, once our participants find out they were tricked, they are interested to learn why they made the decisions they did.

It's through these real-world-type experiments that we can find out what leads people to hurt others or help them, to break the rules or honor them, to seek revenge or take the high road, to mate for life or have a series of one-night stands, and so on. We've designed experiments that test everything from whether people will cheat one another to whether they'll punish straying partners, help someone who got screwed over, show prejudice toward groups they don't even know, step up as a leader, or act like a hypocrite. What's more, each of our experiments, in one way or another, is designed to reveal something about not only *what* people will do but also *why* they do it. In this book, we'll welcome you into our lab and invite you to tag along with us as we conduct these experiments and others. Along the way, we'll share with you what we and other researchers have found about the workings of the social mind. We intend to leave you not only with a new understanding of why you do what you do and how your "character" works, but also with scientifically tested strategies for gaining some control over it.

2 / Hypocrisy vs. Morality

Why no one should throw stones

It was the eve of Valentine's Day, 2008, when George slipped out the side door of one of Washington, D.C.'s most luxurious hotels. All the pieces for the night's romantic rendezvous were in place—he had secured a lavish suite, arranged for his lover's ride to the encounter, and made sure the champagne was on ice. He had even carved out several hours for this tryst, which, for a man of his stature, attested to its importance. George was a powerful man of powerful means. He'd spent the majority of his career in noble pursuits, fighting depravity and corruption of every type, protecting the little guy at every turn. George was under a lot of pressure; tonight, he told himself, he deserved a night off.

As he entered the grand lobby of the hotel and headed toward the elevator, his pulse quickened in anticipation of the romantic pleasures that awaited him. But George Fox, as Eliot Spitzer preferred to be called when he checked into the Mayflower Hotel, wasn't going to meet his wife. No, that night Governor Spitzer, who himself had famously crusaded against the scourge of prostitution in New York, working tirelessly to put hundreds of johns behind bars, was in fact a john himself, and he was about to be publicly outed in a major scandal that would destroy both his image and his career virtually overnight.

What's more, that night at the Mayflower wasn't one single dalliance, one isolated moral lapse. No, this anti-prostitution poster boy was a regular client of the Emperor's Club and had spent many hours—and thousands of dollars—in the company of the highest-class call girls. Here was a man who had made ethics and integrity the hallmarks of his administration, a man who loudly and repeatedly decried the decline of good old American family values. Yet Eliot Spitzer (or "Client #9," as he was to become known) would in one month's time be implicated in the most famous prostitution case of the decade and immortalized in history books as the very paragon of moral hypocrisy.

Of course, Spitzer is hardly an anomaly. In our society, examples of hypocrisy abound. Consider how Rush Limbaugh railed against the moral failings of drug abusers while he just happened to be racking up an impressive collection of illegal prescriptions to feed his oxycodone habit. Or how Senator Larry Craig, who very publicly admonished President Bill Clinton for being a "bad boy, a naughty boy," during the Monica Lewinsky scandal, was caught soliciting sexual favors in men's restroom stalls (and, by the way, he was a fierce opponent of gay rights as well).[1] And it's not just politicians. Think about how countless sports icons, from Mark McGwire to Barry Bonds, Marion Jones, and others, have condemned fellow athletes for the use of performance-enhancing drugs, only to later be implicated in juicing scandals themselves. Or how William Bennett, probably one of the best-known advocates for moral education in this country, a pundit who repeatedly and vocally extolled the benefits of self-control and restraint in his best-selling tome *The Book of Virtues*, was, during the many years he spent promulgating this message, a gambler extraordinaire. While his political organization, Empower America, was publishing editorials decrying lawmakers who "pollute our society with a slot machine on every corner," he was playing stakes so high that he gambled up to $1.4 million in a single two-month period.[2]

As each one of these people fell from grace quickly and publicly, most of us couldn't help wondering what they had been thinking. How could they have been such hypocrites? How could they have done the exact opposite of what they proclaimed to be virtuous behavior? These are all good questions, and they've been exhaustively debated. But they're the wrong ones to ask. It's not that these people ignored or purposely defied what they thought was right. No, it's that what they thought was right was relative. As we'll show in this chapter, hypocrisy isn't so much a matter of violating your own moral beliefs as it is of shifting your moral beliefs to suit your needs and desires at any given point in time. So the right question isn't whether Spitzer and the rest knew what they were doing was wrong. Rather, we should ask how their minds tricked them into believing, at that particular moment, that what they were doing was okay.

Now, you may still be thinking, "But everyone knows politicians and celebrities are an exceptionally questionable lot when it comes to morals. They're not like the rest of us good folks. *We* certainly would never act like that, would we?" Well, that question raises an interesting point. Is hypocrisy a trait confined to a few bad seeds? Or might the potential to act hypocritically lurk in all of us? Given our theoretical view, we suspected the later. Not because we believe human beings are inherently flawed or morally bankrupt but because, as we discussed in Chapter 1, the mind is subject to a constant and often hidden battle that frequently drives us to say or do one thing one minute, only to turn around and do the very opposite the next.

But how exactly does this battle play out? How can we experience such powerful and seismic swings in our beliefs about right and wrong? This is exactly what we were dying to find out. But there was just one problem: how to study hypocrisy in the lab. Clearly, we couldn't just ask people whether or not they would violate their beliefs in a given situation; after all, no one thinks he or she is a hypocrite, and even if some people did, we sincerely doubted they'd be willing to admit it. No, we needed to create situations where people would have something to gain by going against their stated values—situations that provided as close an approximation of a real-world moral dilemma, with all its true temptations, as possible. So we did what we always do: we staged a situation to put people's moral calculus to the test, to see how'd they actually behave when push came to shove. In essence, we conned them. Hey, it's all in the name of science.

But there was one more complication: to study hypocrisy, we had to see not only how people would evaluate their own behavior but also how they would evaluate the same behaviors when they were committed by others. This meant we needed a "bad guy," an accomplice (or what psychologists call a confederate) we could count on to do something morally questionable so that we could see how the true participants would react. Enter Alex. Alex was one of those fine students who was so intrinsically interested in the workings of the human mind that he was willing to risk the wrath of his peers by acting as the universal jerk (for lack of a better word) in our studies. He agreed to repeatedly screw over other students and let them judge him for it. Now that takes guts!

Meet your inner hypocrite

"Maybe I *can* get out of here early," James thought as Carlo Valdesolo left him alone in the lab. James was there to take part in an experiment that he believed (okay, because we told him so) was examining problem-solving skills. When James arrived, Carlo sat him at a computer and told him that he would need to complete one of two tasks. One was a fun and easy photo hunt that would take only about ten minutes. The other task was a series of logic problems that Carlo warned might be difficult and might take as long as forty-five minutes to complete. But, as Carlo next explained, he, as the experimenter, needed to be kept "blind" as to which task James and the other participants would complete so that he wouldn't bias their performance in any way (a false but believable tale; you'll see why we needed it in a minute). "So," Carlo went on, "certain participants are going to be randomly selected to assign themselves—and, therefore, the person going after them—to one of the two tasks. The tasks alternate, so the next person will complete whatever task the first person doesn't." James just happened to be one of these "deciders." (In reality, of course, all our subjects were "deciders.") Next Carlo casually told James that most people believe the fairest way to make a choice is to flip a coin, and handed over a computerized device that flipped a virtual coin, "just in case you want to use it." Then Carlo left.

Now came the fun part (for us, that is): showtime on the hidden cameras. James sat back in his seat, looked at the coin flipper, looked back at his computer screen, and did what a whopping 92 percent of his fellow participants would also do—assigned himself to the quick, easy task without using the flipper. And in so doing, he knowingly doomed the next soul to forty-five minutes of drudgery. Then, just as James finished the short task, the computer posed the following question to him (which, of course, was the point of the whole experiment, even though he didn't know it): "How fairly, on a scale ranging from *not at all* to *very much*, did you act in the assignment procedure?"

It's a simple but telling question, as it requires people to evaluate the rightness of their actions on a very fundamental dimension—fairness. When we tallied the results, we found that the people who assigned themselves the easy task, like James, rated their actions on average somewhere near the middle—they believed their behavior to be not completely fair but not terribly egregious either. Simply put, they believed taking the easy task at someone else's expense was a somewhat acceptable thing to do.[3]

"Okay," you might be wondering, "so what? Maybe most people don't see this behavior as such a bad thing. That doesn't make them hypocrites." But wait, we weren't done yet. Soon it was Jack's turn. Jack also was there to take part in a study that was purportedly about problem solving. This time, however, we made one important change to the experiment. Carlo told Jack that he wouldn't be solving any problems. Instead, his job was to provide feedback on the experiment and problem-solving tasks as an observer. Jack, then, was to surreptitiously watch (via webcam on his computer) as another person went through that same procedure James had just completed. That meant he'd be able to see and hear everything that happened in the session, including whether the person flipped the virtual coin or just took the easy task for him- or herself. Then Jack would be asked his opinions about the whole process. Simple enough.

Jack readily agreed to participate, enjoying the idea of playing the somewhat stealthy role of the "secret watcher." At this point, Alex, our universal "bad guy," entered the room. Jack watched and listened as Alex received his instructions from Carlo. They were the same as before. Alex was told about the two tasks, and that he was selected to be the decider. He was presented with the virtual coin flipper and then left alone. Jack then watched as Alex looked at the flipping device, shook his head slightly, turned back to his computer, and assigned himself the preferable task. Next Jack's computer stopped showing what Alex was doing in the other room and asked for Jack's feedback on the experimental procedures, including his opinion of how fairly Alex acted. This part of the experiment was repeated forty-five more times, all with different "Jacks."

In this version, ratings were not so charitable. Jack and the other "watchers" universally condemned Alex for choosing the good option for himself. To them, the decision was completely unfair and immoral, and even colored their opinion about poor Alex himself. Jack wasn't the only one who gave Alex a dirty look when passing him in the hallway after the experiment; one woman even stopped to lean in, look disparagingly at him, and whisper, "I know what you did." Alex was shunned, a moral outcast. Good thing for him he was graduating soon.

Now, remember, in both of the situations we posed, the same decision occurred: one person chose to assign himself the preferable task at another's expense rather than risking a coin flip. The only difference is who was judging the choice: the person who made it or an outside observer. Yet that was enough to produce wildly different answers to the question of fairness. If the scales of morality were fixed, this shouldn't happen—the answers should be the same regardless of whether people were judging themselves or someone else. An act of cheating should be dishonest, an act of selfishness should be selfish, no matter who committed it. The "badness" of a transgression shouldn't depend on the identity of the transgressor, right? But this is not what happened. People judged the selfish act as far less morally reprehensible when they committed it than when someone else (Alex) did. And it wasn't that one group simply had higher moral standards than the other—we assigned students to the two conditions randomly, as we do in all our experiments, to control for this type of complication. Here, then, we had the very picture of hypocrisy, among the most normal of people.

Now, it's true that sticking someone with thirty-five extra minutes of work isn't exactly a sin on the scale of cheating on one's wife with a high-class hooker. Still, these results tell us a great deal about the nature of hypocrisy and why it's so easy for any of us to fall into its grip. First, they show that our judgments of what is a morally acceptable action seem to be quite fluid. Second, they tell us that our short-term impulses for rewards in the moment—whether those rewards are a night of uninhibited passion with a stranger or getting out of a tedious lab experiment in time for happy hour—can temporarily squelch the voice reminding us about the benefits of a solid reputation in the long term. It's not that we silence this voice purposely, or even consciously; it's a result of the ongoing battle we've been talking about between our short-term interests and our long-term ones. When we act hypocritically, then, it's often not that we're ignoring or deliberately disregarding our beliefs and morals; it's merely that our short-term concerns have momentarily triumphed. That's exactly what happened in this experiment. The people who judged themselves more leniently for taking the easy task weren't aware that they were allowing their minds to adjust their beliefs about right and wrong to serve their immediate interests. It's just that when our inner grasshopper—our desire for short-term rewards—wins us over, we're very good at rationalizing our actions, tricking ourselves into believing that what we did wasn't wrong.

At this point, you may be wondering what happened to the mental mechanisms of the ant, the ones that are supposed to protect us from being socially ostracized by steering us toward fairness and honesty. We had the same question. If hypocrisy were allowed to run completely unfettered, how could we ever trust anyone's judgments or even our own? Selfishness would reign, stable relationships would be impossible to sustain, and our social order would essentially fall apart. So the mechanisms of the ant must be working to some extent, trying to put the brakes on shortsighted, self-serving judgments. In the case of hypocrisy, we figured those mechanisms would look a lot like guilt. The problem, though, was that with the current experiment, we couldn't tell whether the desire to avoid the unpleasant task had trumped the guilt or whether those students simply hadn't felt any guilt at all. To answer this question, we had to go back to the lab.

As we noted in Chapter 1, every decision we make in our lives involves a whole host of related mental processes; some we control and others we don't. And because so many of these processes lie beneath our level of awareness, disentangling them can be a bit tricky. Still, if we wanted to uncover the actual workings of the social mind, we needed a way to isolate the systems we control from the ones we don't. If our theories were correct, hypocrisy would be, in part at least, a function of time. We suspected that at the outset of our experiment, our participants would feel some innate, automatic impulse to be fair, especially given the long-standing importance of fairness norms for interpersonal relationships. With every passing second, though, each person's grasshopper would work harder and harder to help him or her rationalize acting unfairly in order to win immediate gains. It would be almost as though, if you listened closely enough, you could hear the grasshopper saying, "The experiment is anonymous. The other person being screwed over wouldn't know what was happening, so there's nothing to lose," as it worked to tip the scale its way. In other words, we suspected that the "hypocrisy" we observed in the experiment resulted from the mental jujitsu involved in this act of rationalization. To test this theory, all we had to do was stop the rationalizing in its tracks.

One common trick psychologists use to disentangle dueling mental processes such as these is simply to inhibit one of them. We figured that if we could hamper or even knock out the rationalizing part of the brain by keeping it busy, then we would be able to see what, if anything, the ant was up to. So we decided to have our participants memorize strings of random digits. After all, we figured, it's more difficult to craft clever justifications when your mind is working hard to remember something.[4] Here's the way it worked. We ran the two conditions of the experiment (judging oneself and judging another) exactly as before, but with a single exception. This time half the students were presented with a different set of seven digits before each question that the computer asked them in the final task, including the question about fairness. They were told that after they answered each question, they would have to type in the seven digits that preceded it.

The point was this: In order to remember the string of digits, the participants would have to mentally rehearse the digits while they answered the question. This kept their minds occupied, so they wouldn't have the mental energy left to devote to rationalizing away their own less than moral actions. In essence, we tied the hands of the short-term system, to see whether or not the long-term one—the one fighting on the side of fairness—was working. As it turned out, it was. When we prevented rationalization by limiting the systems of the grasshopper, the hypocrisy we'd observed earlier completely vanished. In a fascinating twist, this time people judged the act of assigning oneself the easy task without using the flipper just as morally objectionable when they committed it as when others did—there was absolutely no difference in how morality was applied.[5] What this finding tells us is that we *do* feel in our gut that screwing over the next guy is wrong. The pangs of guilt are immediately there at the intuitive level; it's just that our minds are very good at squashing them with reasoned excuses when it serves our short-term interests, especially when it's unlikely that we'll be caught.

So if the desire to avoid a mere thirty-five extra minutes of tedium was enough for most people's rational minds to overrule their intuitive drives to be fair, it suddenly doesn't seem so surprising that Limbaugh was able to rationalize popping oxycodone while condemning "drug addicts," or that Spitzer told himself it was okay to plan trysts with employees of the Emperor's Club while fighting prostitution, or that so many athletes think it's fine to use steroids to help them win medals while at the same time decrying the problem. After all, you can't deny that the short-term rewards of all these activities are very seductive; the incentives to rationalize away any moral qualms about behavior are all there. What's more, as our studies and others show, when an incentive to commit an immoral act is salient, our rational minds are very good at coming up with reasons to justify it. For Spitzer, maybe it was that the pressures of the job and power that came with it entitled him to some extramarital pleasure. For those doping athletes, it might have been that the bump in pay they'd get from winning a game, a series, or a title would help them better provide for their families. As for Limbaugh and Bennett, well, addicts are the best of all at this game. Point is, the excuses our minds can come up with are many and varied. And when we try hard enough, we can convince ourselves of any of them. Considering that the same potential for hypocrisy resides in each of us, it suddenly seems a lot more perilous to start throwing stones.

The dynamics of elastic morality

The fact that hypocrisy can come so easily to any of us goes to show just how elastic our ethics and morality can be. It's not that we don't have any deeply held ideas or values about what is right and wrong. It's just that these basic notions are malleable and subject to change at times. The tricky part of acting morally, then, doesn't center on if we can judge what's right or wrong and act accordingly—it centers on how we judge right and wrong and on how changeable these judgments, and thereby our character, can be.

The above evidence shows pretty conclusively that our moral codes aren't completely stable or static and can change from one situation to another. What you may not realize, however, is that sometimes they can change even for what appears to be no reason at all. Over the past decade, much research has begun to show that our morals are often shaped as much, or even more, by our emotional responses than by our so-called rational ones. Don't believe us? Consider this example that our colleague, the psychologist Jonathan Haidt, often poses to participants in experiments:

> Julie and Mark are brother and sister. They are traveling together in France on summer vacation from college. One night they are staying alone in a cabin near the beach. They decide that it would be interesting and fun if they tried making love. At the very least it would be a new experience for each of them. Julie was already taking birth control pills, but Mark uses a condom too, just to be safe. They both enjoy making love, but they decide not to do it again. They keep that night as a special secret, which makes them feel even closer to each other. Was it okay for Mark and Julie to make love?

Assuming you're like most people, the answer is probably a resounding no, likely accompanied by an uncomfortable visceral feeling and a look of disgust. After all, everyone knows incest is unequivocally immoral and can only lead to trouble. Yet when people are asked to explain their rationale for *why* Julie and Mark's action was so repugnant, something interesting happens. Some cite the health risks associated with inbreeding, only to be reminded that the multiple forms of birth control the siblings used preclude this possibility. Others venture that the act surely will cause psychological harm to one or both of the siblings or destroy the dynamics of the family, but Haidt reminds them that the scenario ruled these possibilities out as well. Usually people keep scratching their heads, searching for a logical explanation to justify their moral outrage, but come up empty. After all, there are no objective consequences of Mark and Julie's actions. Still, most people steadfastly maintain that this act is just *wrong*, even though they can't seem to articulate exactly why it's wrong in the current case. Even when presented with all the reasons why no harm could possibly come from this night of lovemaking, their gut aversion to the incestuous act is so powerful, they can't shake it.

The reason this gut feeling is so strong is because it serves an important evolutionary purpose. Thousands of years ago, these innate emotional responses were all our ancestors had to guide social living. Remember that because of the way our brains evolved, the mental capacities for abstract reasoning about things like ethics constitute a relatively new ability, evolutionarily speaking. The capacity for emotion, however, is much older and existed well before we had the cognitive wherewithal to weigh the consequences of our actions. So way back when, before the capacity for reason evolved, our innate revulsion to incest was very adaptive; it protected our species from serious risk, namely, the genetic defects and diseases associated with inbreeding. Back then, sex without the risk of procreation didn't exist. Trojan wasn't always around to help prevent unwanted pregnancies, and oral contraception is less than a century old. So the moral revulsion you feel at the thought of sex with a sibling was, in a way, an ancient form of birth control. After all, it's difficult to be too turned on when you're feeling disgusted.

The point is, emotions regulated our ancestors' social behavior and still perform this function for us today. Emotions tell us quickly and almost effortlessly what we should do in a given situation. Sometimes, as with incest, there is only one right answer. Even if a small physical pleasure were to come from the act itself, we know instinctively that it would pale in comparison to the biological risks, so the long-term system (the ant) wins easily. However, other times, as we just saw in the hypocrisy example, when both short- and long-term benefits exist (e.g., getting what we want now vs. building a reputation as fair and trustworthy), the battle of our emotions may have no clear winner but rather may shift back and forth.

It's important to realize, though, that finding the best course of action when moral dilemmas arise isn't always as simple as choosing to go with your gut. Even though it's tempting for many of us to want to trust our intuition, especially in light of having just seen how so-called rational thought can lead to hypocrisy, it's becoming clearer than ever that when it comes to moral decisions, there is no perfect strategy. Following your gut is no more foolproof than listening to reason. The following experiment illustrates why.

Derailing the moral mind

John and Ben had signed up to take part in an experiment about opinions. It sounded interesting; who doesn't like to tell others what they think? "We're trying to understand your views on the world," Carlo told them when they arrived at the lab. "So you'll simply see some questions appear on your computer screens, and all you need to do is tell us which of the presented options you think is the right or most acceptable one." Piece of cake, thought John and Ben. Carlo continued, "We're also collecting opinions on television shows, so before we get to the main questions, we're going to have you watch a brief video clip and tell us what you think of it." For John and Ben, it didn't get any better than this. They were actually going to get paid to watch television.

Carlo sat John and Ben at their computers, gave them headphones, and ran the video clips. When the clips had finished, hypothetical situations began popping up on their screens. Most of the dilemmas were quite innocuous and neutral, such as "If you were marooned on a desert island, would you rather have a jar of peanut butter or a book of matches?" But, unbeknownst to John and Ben, the third scenario was the one that mattered:

> You find yourself standing on a footbridge overlooking trolley tracks. Barreling down these tracks is a runaway trolley that, if allowed to continue unimpeded, will run over and kill five workmen who are up ahead on the tracks. Standing next to you is a rather large man. The only way to stop the trolley would be to push the large stranger off the bridge and onto the tracks, whereby it would kill him but stop the trolley before it reached and killed the other five. Should you push him?

On this particular day, John answered quickly and decisively: no. For Ben, the answer was easy as well: yes. Exact same moral conundrum, completely opposite answers (and it's not that John and Ben differed in age, background, or some other fundamental way—we'd controlled for that). Why the difference? The answer lies in that video clip.

It just so happens that the clip Ben watched right before he was presented with the moral dilemma was a *Saturday Night Live* comedy sketch. John, on the other hand, got stuck watching part of some dull documentary on life in remote Spanish villages. "So what?" you're probably wondering. "Something as minor as watching television couldn't possibly lead our intrepid participants to report such wildly different views about whether it's morally acceptable to push someone to certain death, could it?"

Actually, yes. In fact, when we analyzed the results of the seventy-nine people like John and Ben that we ran through the experiment, those who watched the *SNL* skit were more than three times as likely to say they'd push the man off the bridge.[6] It seems perplexing at first. After all, these clips didn't have anything to do with weighty issues of morality or life and death. But that's exactly the point. Watching these videos didn't change people's deeply held moral principles or beliefs. It *did*, however, change their emotional states, and that's what matters. As we've said, our emotional instincts and impulses often guide our moral choices. Consequently, anything that can alter what we're feeling has the potential to derail (no pun intended) our moral reasoning, whether we're aware of it or not.

Turns out this is exactly what was going on with Ben and the others like him who decided to push the stranger off the bridge. They weren't callous, coldhearted killers. Nor were they simply unfeeling logicians. It's not that their characters were fundamentally different from those of John and the others like him; it's just that their intuitive feelings got smacked down due to a little experimental interference.

Decades of research have shown that when we're experiencing an emotion, it can't help coloring all our actions and decisions—even ones that have nothing to do with what we're feeling in the first place. You have a fight with your boss, and you come home and feel like kicking your dog (though we hope you don't). You feel anxious about a new promotion and suddenly think your odds of contracting cancer are higher. Simply put, we all unwittingly use our emotional states as information, or cues, to guide our decisions about what's likely to happen or what we should do.[7] If we're feeling sad, we can't help feeling that depressing things must be just around the corner.

In the present case, those who watched the *SNL* skit were

understandably feeling more buoyant and cheerful than those who had watched the snore-inducing documentary. As a result, the visceral negative feelings that otherwise would have been triggered by the thought of pushing another to his death were momentarily blocked. With these gut feelings held at bay, it became easier to rationally weigh the consequences of the two scenarios and conclude, quite logically, that it is morally acceptable to sacrifice one life to save five others.

To fully understand the role that emotions play in this kind of moral decision, consider what happens if we slightly change the specifics of the dilemma. In this new scenario, the runaway trolley is still barreling down the tracks. This time, however, there are two directions it can go. If left as is, the trolley will roll straight ahead and kill those five track workers. However, if you flip a rail switch, the trolley will be diverted onto a different track, where it would kill only one worker instead of five. Would this change your decision? You're still deciding whether to sacrifice one person's life to save five others, only now you don't have to physically push someone onto the tracks. Would you flip the switch? In our experiment and those of many others, the answer is almost unanimously yes. Flipping the switch is judged the right thing to do. Saving five is better than saving one, period.

If that's the case, though, then why do countless studies reveal that when confronted with the otherwise equivalent version where you have to physically knock someone off the footbridge to save the five others, the vast majority of people (assuming they haven't just been made to feel happy)—a staggering 90 percent—believe it wrong to do so? Logically, it's the same trade-off in numbers saved and killed. The answer, however, has nothing to do with logic. It's much simpler: the two situations *feel* different. Take a moment to think of how it would feel to wrap your hands around the flesh of another living, breathing human as he teeters perilously at the edge of a high bridge, to see the fear in that person's eyes as he struggles fruitlessly to escape your grip. Assuming you don't have psychopathic tendencies and aren't smiling right about now, that pit you feel in your gut when thinking about shoving the guy, even to save five others, results from the intuitive systems of the ant screaming, "Don't do it!" For most of us, this impulse usually wins.

Human minds are programmed to have an innate aversion to inflicting harm on another (unless the person poses a threat), and it is precisely this aversion or sense of horror that usually prevents most people from choosing to push the stranger off the footbridge, even though it might make logical sense to do so. In this instance, the systems of the ant, on the intuitive level at least, are on steroids because, evolutionarily speaking, causing intentional harm to an innocent person is a big no-no. Hurting others outside of war is almost never good for a person's reputation and thus threatens our long-term survival. So considering pushing someone off a bridge, even for a good reason, makes us feel quite uneasy. However, in the switch version of the dilemma, although the trade-off between life and death is quantitatively the same, the action in question doesn't alarm the ant to the same degree. Imagining throwing a switch doesn't feel nearly as awful on a gut level as pushing someone to his demise, even if the results are the same.

Recent research in neuroimaging supports the view that the decision about whether or not to actively push the man off the bridge is guided by intuitive emotional responses, whereas the decision about whether or not to flip the switch is more grounded in conscious reasoning. In groundbreaking work, psychologist Joshua Greene and his colleagues used fMRI techniques to peer into people's brains as they grappled with these moral decisions. They found that the centers involved in experiencing emotion were much more active when people were considering whether to push someone off a footbridge than they were when the question was whether to flip a switch.[8] In the case of the footbridge, the ant pushes hard on the intuitive level to keep us from pushing hard on the large stranger. In the case of the switch, there is no initial intuitive response, and so the rational mind doesn't need to fight against an initial decision. As we said earlier, sometimes the choices of the intuitive and rational minds can differ even when the goals are the same. Because intuitive mechanisms are guided by what has tended to work best over millennia (e.g., don't directly harm someone), they can short-circuit when confronting novel situations that our ancestors never faced. Back then, if you were going to kill someone, you had to do it with your own two hands; there were no switches.

Given that we're clearly not on the savannah anymore, this raises another set of questions: Can't the mind adapt? Are we doomed to forever make decisions that feel right but end up being logically or even morally wrong? Well, let's go back to our first experiment, with the video clips. If we look at the relatively few people among those who watched the comedy clip who *did* decide to push the guy off the footbridge, an intriguing pattern emerges—they took markedly longer to make their decision than did the majority of people who chose not to push the hapless stranger. In this finding, you can see the tug-of-war between the intuitive and rational minds. The reason the decision to push the one to save the five others took longer to make was precisely because people's minds had to work to override their intuitive impulse not to cause direct harm to someone. In essence, their minds were doing exactly what the minds of Eliot Spitzer, Rush Limbaugh, and all the other "hypocrites" were doing: constructing rational explanations for their actions and decisions. But there is one fundamental difference: unlike the hypocrisy cases, the trolley dilemmas don't present any immediate potential for self-interest, so reasoning can be more objective. Without anything to gain in the short term by making one decision or the other, the grasshopper doesn't perk up to fight the ant.

The significance of these experiments is twofold. First, these findings unequivocally show that what we feel, not only what we think, guides our moral judgments. Second, given that our feelings can and do change quickly and seemingly unpredictably, our moral judgments, and therefore our character, are quite flexible too. The mechanisms of the mind aren't perfect. Though they serve us well most of the time, they can be tripped up by context. Potentially more troubling still is that such changes in context aren't always random; they are readily susceptible to intentional manipulation. After all, if something as seemingly trivial as watching a short video clip or hearing a joke can alter our moral judgments, imagine how vulnerable we are to deliberate manipulation by politicians, lawyers, PR specialists, ex-boyfriends, and others who try to shape our views about right and wrong, or guilt and innocence, by playing on our feelings. When our scales of morality are as wobbly as we now know them to be, it can be incredibly easy for other people to deliberately tip them.

The perils of dirty tissues and soapy hands

If simply watching a television show can alter your morals, where does the power of emotions stop? Surprisingly, there really isn't a good answer to this question. Basically, anything that can appeal to your intuitions and change your feelings can pretty much impact your moral decisions. Take, for example, a dirty tissue—the crumpled kind oozing with some bodily fluid you'd really rather not think about. What's the first feeling that popped into your mind when you pictured this image? If you're like most people, it was queasiness or a feeling of disgust. Okay, you may be thinking, "So what? A used tissue is gross." We agree. Such an object repulses us—the feeling stems from deep down in our gut. Funny thing is, though, sitting next to a used tissue can actually sway your moral judgments about completely unrelated issues, such as gay marriage or failing to recycle. Why? Because that feeling of disgust can give one side of the scale a head start in shaping your judgments.

Simone Schnall and her colleagues demonstrated just this fact.[9] In one series of experiments, they asked participants to rate the moral acceptability of various acts: How immoral is it for first cousins to have sex? To eat your dog after it dies? To eat your friends if they're killed in a plane crash that leaves you stranded on a glacier? But unbeknownst to the participants, the researchers had "decorated" the room where these decisions would be made (for half the participants, that is) prior to their arrival. This lucky half found the room to be, shall we say, a little messy. The researchers replaced the clean chairs with stained ones. They replaced new pens with chewed pens. They replaced empty trash cans with filled ones, topped off with dirty tissues. And lo and behold, the participants who made their decisions in the messy room overwhelmingly rated each possible moral transgression as far more reprehensible than did their counterparts in the clean condition. Why? Because the feelings of disgust generated by the mess primed the intuitive system to be disgusted by whatever happened to come next. In essence, that feeling of disgust bled over onto the next things that entered consciousness.

So when people were asked how they felt about a somewhat tenuous moral action, the answer was already there: it was disgusting. And condemn those actions they did.

Luckily (or maybe not, depending on one's point of view), we can sometimes use this vulnerability of the mind to our advantage. Take for example, the case of Sam. Sam was a friend of one of ours in college. (Okay, his real name isn't Sam, and no, we won't tell you which one of us knew him. We have to give the guy some cover!) Anyway, Sam was a nice guy from New York City who suddenly arrived at college as a freshman and realized that he could reinvent himself. To put it simply, Sam, who had never had much luck with the ladies, became a player. As the weeks of the fall semester passed, Sam's friends couldn't help enviously noticing that he was dating more and more women. (Well, it's hard to call it dating when the relationships usually consisted of one-night stands, but let's go with it.) Women were drawn to Sam because, amazingly, even though he was playing the field, his reputation was still that of Mr. Nice Guy, someone who would respect you in the morning and be there when you needed him. How was he fooling them? To his buddies, he seemed to have become a total playboy, some sort of modern-day Lothario racking up notches on his bedpost. But to the ladies, he was seen as sensitive, caring, and sweet. It was puzzling. Then his friends noticed one thing: Sam seemed to have developed a new habit of stopping off at the restroom at frequent intervals to wash his hands. Not to use the toilet or look in the mirror, just to wash.

Now, Sam was no clean freak—far from it. His room was as untidy as ever. He still lived for days in the same pair of jeans. He wasn't shaving and getting haircuts more often. And he certainly wasn't trying to avoid germs—he'd take a drink from anyone's glass. The next year, by which time he'd settled into a long-term relationship with a woman he'd met over the summer, the hand-washing behavior stopped as abruptly as it had begun. And so the mystery lingered.

It wasn't until years later that we found the answer. Sam's change in character had simply been a temporary victory of the mental system favoring his short-term interests (his desire for casual sex). And the hand washing? That was simply a subconscious attempt to assuage his own feelings of guilt about using these women. Just like Lady Macbeth, he was trying to wash his sins away. He didn't know it, but in adopting this one little ritual of cleanliness, he was alleviating feelings of disgust and guilt at his less than upstanding actions. And it worked. Once Sam had convinced himself he was still the same good guy he had always been, he projected that image to the women, who in turn were readily convinced.

The science underlying this "Macbeth effect" has been documented by Chen-Bo Zhong and Katie Liljenquist in a series of clever experiments.[10] In one, Zhong and Liljenquist found that participants asked to recall an unethical deed or write about an unethical act later purchased more cleaning products than their guilt-free counterparts—their intuitive minds felt a need to be "clean." Even more pertinent to Sam's case, Zhong and Liljenquist found that if they allowed guilt-racked participants to wash their hands after recalling their questionable actions, the need to "cleanse" themselves, or atone for their sins, went away. Among those with a guilty conscience who were allowed to wash, fewer than half as many volunteered to help a peer in need of assistance. Just as dirty tissues prime us to feel morally repulsed, the simple act of washing—the feeling of being clean—sends a signal to the older, intuitive mental mechanisms that moral violations have disappeared. Thus it's easier for the "sinning" to continue.

Like all the other emotional impulses we've discussed in this chapter, the feeling of disgust has an important evolutionary purpose. It began as a simple reflexive feeling and action meant to keep our ancestors away from dangerous things. Think about it. Eating rotten meat, feces, or toxins is certainly bad for you, and consequently all are considered disgusting. Over thousands of years of cultural evolution, that original biological disgust response came to be generalized not just to impure food but to all things considered "impure." This is why feelings of moral disgust or guilt can be held at bay through simple acts of physical cleansing. On an intuitive level, feeling clean is feeling clean.

Sinful saints or saintly sinners?

Are we all hypocrites, then? Are all our moral compasses broken? Do we even have compasses to begin with? The answer to these questions, it seems, is both yes and no. We all *can* be hypocrites, but we're not *always* hypocrites. Acting hypocritically is different from being a hypocrite. Sinning is different from being a sinner. The first implies an instance; the second suggests a deep-seated disposition. As we've said, our moral compass isn't broken, but it isn't fixed either. It just works differently than most people think it does. As our research reveals, not only is our morality flexible, but the scale that determines it is constantly being tipped back and forth by mechanisms that operate under our radar.

Don't feel bad about this news. It doesn't make us inherently flawed, weak, or bad people. It's not that we don't feel pangs of guilt over our own morally questionable actions; it's just that our minds are remarkably good at quieting them. Even Spitzer probably felt pangs of guilt as he made calls to the Emperor's Club. But then his desire for immediate pleasure ("I need some fun and those Emperor's Club women are so hot") went to battle with the voices warning him about the long-term consequences ("This can only spell problems for my family and my career"). And, well, we know which of them won.

Spitzer is no different from the rest of us in terms of the way his mind works. Whether it's because of the battle between our own inner mental mechanisms or changes in our external environments, we don't always act as morally as we'd like. But that doesn't mean we should give up trying. Understanding how the system truly works is the first step toward being able to manage it better.

For example, now that you know how readily moral judgments are influenced by emotional states, it becomes easier to understand why telling that off-color joke about your mother-in-law seemed okay yesterday at a celebratory dinner but feels like a horrible idea today. Why it seemed okay last night to sleep with that married person you met at happy hour, even though you woke up this morning deeply regretting it.

So how can we avoid falling prey to such lapses in moral judgment? The first step is to remind ourselves that if we're feeling happy or aroused, whether it's because we've been imbibing or just because we've been having a good time, those feelings can color moral judgments by squashing the emotional impulses of the ant—those that are looking out for our long-term interests—by giving precedence to impulses favoring pleasure in the here and now. So when you're laughing or partying it up, it helps to realize that the warm glow you're feeling may be blocking out the hesitation you normally might have felt before doing something you're likely to regret the next morning. Of course, we've seen it can work the other way so, don't ignore it. *Feel* it! Forget anything you've read about the importance of reason in making good decisions. If you're feeling a visceral emotion, weigh that feeling in your conscious analysis of what to do. Of course, it's not the only piece of information, but it's an important one. It also doesn't mean that emotions will always be right; as we've seen, many gut emotions stem from an ancient calculus that no longer applies (remember the footbridge dilemma), while others are colored by the situation you're in. The point is not to trust either your conscious will or your intuitions 100 percent of the time, but to try to see whether what you're feeling and what you're thinking stem from ulterior motives or extraneous contexts.

Lastly, don't assume you're good at this tactic or that you'll get it right every time. As we've seen, none of us is a saint; we all err in our moral judgments every now and again. And in fact, a little humility can be useful. As recent work by Sonya Sachdeva, Rumen Iliev, and Douglas Medin at Northwestern University has shown, having an outsized sense of moral superiority often gives people license to act less morally in the future.[11] The researchers asked participants to use one of two sets of words in writing a short story about themselves—a set of words suggesting high moral character (e.g., *generous*, *caring*) or one suggesting low moral character (e.g., *greedy*, *disloyal*). After a little time passed, they asked the participants if they'd like to make a donation to charity. What they found is not only at odds with what most people would expect but opposite to the view of fixed character as well. The people who wrote stories about themselves using the "moral" descriptors gave far *less* on average ($1.11) than did their counterparts who used the immoral descriptors ($5.56). Describing oneself as moral didn't make these people act morally. To the contrary, trumpeting their moral qualities apparently gave their short-term systems greater room to urge them to keep more money for themselves. As we said, the fight between the grasshopper and the ant isn't usually a fair one.

It's easy to see this same phenomenon outside the lab as well. Take, for example, Oral Suer, the former CEO of the Washington, D.C.–area United Way. He labored tirelessly over his thirty-year career to raise more than $1 billion for local charities, but it was later revealed that he had been diverting hundreds of thousands of dollars from it to "reward" himself for his charitable work. The same phenomenon may also have been partially at play in Spitzer's decisions to indulge himself. After all, didn't all his victories against the scourge of corruption give him license on some level to enjoy himself in an unsavory act now and again? The point here is to be careful by knowing where these pitfalls lie. The human mind, as we'll continue to see, is capable of much contradiction and all manners of tricks.

3 / Soul Mate or Playmate?

What makes Mr. or Ms. Right go wrong

Lust and love. We want them both, and the right combination can make life worth living. Most of us think we can tell a lot about a person just by the balance he or she strikes. Gentleman or playboy? Matron or cougar? Flirt or slut? These can be fine lines; after all, it's one thing to have a reputation as a Don Juan, but it's quite another to be seen as a Roman Polanski. Show up on someone's lawn and serenade him or her with ballads, it's romantic. Show up five nights in a row, it's grounds for a restraining order. So what determines whether we end up behind the bedroom door or out in the cold? Celebrating a fiftieth anniversary or a fiftieth notch in the bedpost? It may seem straightforward—lust is one thing, true love another. The former is a sign of vice, the latter a marker of virtue, right? Not exactly. The real story, as you might suspect, is much more complicated.

Consider the case of Tiger Woods, who has suffered one of the most rapid and monumental falls from grace of any public figure in recent memory. Woods had erected a billion-dollar empire on a foundation of a sterling public image (well, that and a great golf game)—that is, until his "car accident" of November 30, 2009, when the web of lies he had spun to conceal his multiple affairs and indiscretions began to unravel. Prostitutes, "VIP clubs," crude text messages to multiple women—a tawdry picture indeed. The weight of the shame subsequently heaped upon him by the media, his sponsors, and not least of all his fans was so great that Woods, once hailed as the greatest golfer of his generation, was forced to take a leave of absence from the game.

All of which left his fans asking the inevitable question: just what kind of person had they been rooting for all these years? Who was the real Tiger? Once thought to be a virtuous family man, as disciplined and devoted a husband and father of two as he was an athlete, overnight he became cast as an insatiable womanizer who'd thrown away both his thriving career and his loving family simply because, to put it crudely, he couldn't keep it in his pants. So which was he, really? If you're following the argument we've laid out so far, you know that Tiger's morally questionable actions stemmed not from some innate, deep-seated flaw in his character but rather from an ongoing battle between dueling forces: his desire for the long-term stability and devotion of Elin Nordegren was pitted against the short-term pleasures of trysts with cocktail waitress Jaimee Grubbs (among others). In other words, it was the voice of the grasshopper steering him toward lust vs. the voice of the ant pushing for love. And by now, nearly everyone in the Western Hemisphere knows which one came out on top.

It is easy to mount our high horses and judge Tiger, to condemn him as a bad person. But as a richer understanding of character emerges, the line between "good" and "bad" begins to blur. While some might consider lust, or the desire for sexual flings, "bad" and the urge to settle down and raise a family "good," the mechanisms underlying both lust and love actually serve equally important purposes. Cheating and philandering, of course, often bring ill. Yet, as we'll learn in this chapter, the impulses behind them can actually prove quite beneficial at times.

The origins of love

To begin to understand how the battle between lust and love plays out, we first need to define our terms. Lust is an easy one—it's the physical attraction that underlies our desire to have sex with another person. But defining "love" is a bit more complicated. Conventional wisdom would have us believe that love involves the search for one's soul mate. This notion goes back as far as Plato's *Symposium*, which explains the origin of love with the following story. Humans, Plato wrote, were once a race of two-headed, four-legged, and four-armed beings. After they dared to threaten the power of the Greek gods, the gods punished them by showering the earth with lightning bolts, severing each person into two halves—that is, into the form of modern humans. Love, as the story goes, is the lifelong pursuit of our "other half."[1]

Centuries of art, literature, and music tell essentially the same tale—that to find true love is to find that one person who "completes us." One only has to read a love poem or watch a romantic movie (especially one of the Meg Ryan variety) to know that love is about finding our soul mate. But how do you know when you've found "the one"? Hollywood would have us believe that you just *know*.

You just know? What a horribly unsatisfactory explanation for one of the most important decisions in life. If we just know, then what about the countless people who fall in love with the wrong person? Who fall for someone and never hear from them again? Or who are smitten with one person one day, then indifferent to that same person the next? People who cheat on their partners, or who invest months, years, decades in a partner who cheats? Did they somehow *not* know? Or did they just get it horribly wrong? We think the problem is not that they were wrong but rather that the common understanding of love is a bit misguided. Forgive us for sounding cynical or unromantic, but as you will learn in this chapter, science has shown that our culture's simplistic, storybook notion of "one true love" is, quite simply, wrong. As our research and the research of others shows, whom we love isn't determined by fate or by divine intervention. It isn't about finding some predestined "other half." Instead, like so many of the choices that shape our character, the decision of whether to love or lust is a delicate one, shaped by the battle between long-term and short-term interests that is being waged within us at any point in time. The grasshopper urges you to slip your wedding ring into your pocket and approach that intriguing stranger across the room; your inner ant is telling you to pay your tab and speed home to tuck in the kids.

What guides this battle? Here's where we can turn to a fascinating body of research that is beginning to unpack the powerful arsenal each side uses to sway our actions. Tempting as it may be to write off the man-eaters and womanizers as inherently bad people and the chaste and faithful as fundamentally virtuous, the calculus underlying both types of behavior, as we'll see, serves an essential purpose. Like Tiger, we all have the potential to be in love one moment and in lust the next. What may be most surprising, though, is that what influences our decision making in any given situation is often up for grabs.

Take me home tonight

What better way to begin a discussion about the battle between love and lust than by considering the phenomenon of the one-night stand? As you work your way through a crowded bar, why does the man at table three immediately catch your eye? The choice may seem arbitrary. After all, with his brown hair, medium build, and button-down shirt, he differs physically in no meaningful way from 90 percent of the other patrons. Yet something about that particular guy gets your pulse racing. What is it that piques your interest, urges you to approach, even if you know you shouldn't? What makes you feel drawn to this stranger as if by a powerful magnet? Let's say you give in to this impulse, one thing leads to another, and before you know it, it's the next morning and you're waking up in an unfamiliar bed. As the events of the night before slowly trickle back into memory, you're likely to have one of two reactions. Option one: you pretend to sleep as you scan the room for an exit, then scamper as quietly as you can toward it before succumbing to the ignominy of the "walk of shame"—the frantic 8:00 a.m. scurry home in last night's clothes. Option two: you give your new acquaintance a gentle waking nuzzle, cuddle for a bit, then exchange information and promises of future encounters, maybe even have breakfast

What determines whether this encounter blooms into the topic of a wedding toast or fades into one of those embarrassing incidents you'd rather forget? How do you know whether this could be the start of something grand or the start of a rumor you wish would go away? Moreover, why do particular people catch our eye, and how predictive is this initial attraction of future relationship success? These outcomes aren't nearly as arbitrary as they seem. In fact, the mind is attuned to an assortment of cues, usually registering below our level of conscious awareness, that have evolved over time to tip our decisions and direct us toward the most desirable mates.

Now, of course, subconscious physical cues aren't the only determinants of attraction. But in situations where we don't have much information about a person and there isn't much of an opportunity to meaningfully interact (read: loud music, crowded party, brief encounter—the hallmarks of a one-night stand), first impressions are all we have to go on. And there is strong evidence to suggest that these first impressions are often shaped by the mental mechanisms not so interested in our long-term well-being but very focused on maximizing pleasure in the here and now. That's the deafening voice screaming, "Forget about that guy you've been dating for the last couple of months. This new guy right here, this is the guy for you." Such a voice may seen maladaptive, but actually it reflects a series of evolutionary impulses that have been particularly beneficial to our species over time. So while this voice might seem to be urging us to act against our better judgment, there are some cases where the most desirable mate may actually be the one we don't have to call the next day.

What's hot and what's not: The whys of attraction

Research show that uploading a photo to a Match.com profile increases the likelihood of being contacted dramatically.[2] This isn't so shocking. After all, tell a friend you're dating someone you met on the Internet and your friend's first question won't be "Do you have the same hobbies and taste in movies?"—it'll be "What does he or she look like?" The importance of physical attraction at the start of a relationship cannot and should not be disputed. Attractive people get more dates and have an easier time finding a mate than unattractive people do (incidentally, they are also given better grades for the same work, earn more money in the workplace, are helped more when in need, and even are more likely to receive leniency in court).[3] Sad, we know, but true. You might balk at the notion that your romantic decisions are driven by such shallow considerations, or you might be tempted to remind us that beauty is only skin deep and that you can't judge a book by its cover. Well, as much as we'd like to think these quaint sentiments are true, they really aren't, at least not completely.

We might not always like to admit it, but physical appearances are actually incredibly powerful in shaping our first impressions. In fact, not only do we judge some very important aspects of a book by its cover, but a book's cover often sends a pretty universal signal about whether it's worth reading or whether it's, for lack of a better word, skimmable. There is, surprisingly, fairly widespread agreement within and across cultures regarding who is and is not attractive, and what's more, there's good reason to believe these judgments may be somewhat biologically determined. For example, research shows that we begin to form opinions about attractiveness at an incredibly young age—far too young for cultural influences or past experiences to factor in. Several experiments have revealed that even infants show a preference for faces that adults rated as attractive (of course, the babies couldn't explicitly say which they found attractive, so attraction was measured by the amount of time the baby spent looking at different pictures; longer gazes meant stronger preference).[4] This evidence suggests that concepts of beauty not only are well agreed upon but emerge very early and automatically. It even works the other way as well. People tend to agree on the relative attractiveness of babies, which is probably why attractive babies universally tend to receive more affection and attention than unattractive babies, even in the hospital nursery.[5]

So exactly what is it that we all seem to be agreeing about? What are the features that draw us inexplicably to others and make them so hard to resist? Do the Jaimee Grubbses of the world have certain physical characteristics that can make us forget all about what's-her-name who cooked the dinner that's currently getting colder and colder on our dining room table at home? The answer, it would seem, is yes. That's because judgments of attraction are in part rooted in a suite of automatic, intuitive cues about the relative health and fitness of potential sexual partners. By fitness we don't mean the tightness of their abs or how fast they can run the mile. We mean fitness in the evolutionary sense—the likelihood that these partners can make strong, healthy babies with us (though of course the two definitions can overlap).

Just as we have evolved a taste for sweets because we have a biological need for glucose, we have evolved a taste for particular features of the body and face associated with evolutionary "health": we find certain physical features to be attractive in another person because they signal to us on an intuitive level that this is a person who would be relatively more successful in passing on healthy genes to future generations. The cues, in essence, signal to us that this is someone who will not only be reproductively successful but also pass on his or her "good genes" to our offspring. At that moment it doesn't really matter if this is a person with whom we have absolutely nothing in common or if acting on this attraction will destroy our current relationship; when our evolutionary impulses take over, these people can be difficult to resist.

The scent of a woman (or man)

So our evolutionary urges to mate with the healthiest-looking person in the room can tempt us to stray. But that isn't the whole picture. What exactly are these subtle cues we're picking up on that tell us another person is healthy and genetically fit? Some of them may surprise you. For example, imagine a line extending down from the middle of your forehead all the way to the floor between your feet. Presumably on each side you'll find one eye, one ear, one arm, one nipple, and so on, each more or less equidistant from that central line. This is what's known as bilateral symmetry, and while we all tend to be basically symmetrical, how perfectly each side mirrors the other varies from person to person in tiny but perceptible amounts. Some have one hip a bit higher than the other, some a slightly out-of-place earlobe, yet others an eyelid that droops a little too low. And while we don't always consciously register these slight "imperfections" in potential mates or partners, we do perceive them on some level. Study after study shows we consistently rate people who have more symmetrical features as being more attractive.

Why? It's not the artistically pleasing nature of the lines that matter here—it's the sex potential! In fact, much research has shown that bilateral symmetry is a good predictor of reproductive success.[6]

Mothers of more symmetrical infants, for example, have been found to suffer fewer infectious diseases during pregnancy.[7] Of course, it's not the symmetry in and of itself that makes the mother more resistant to infection, it's just that symmetry is a marker for better overall health. This is equally true for adults, which is why the less someone's earlobe droops or the more evenly spaced someone's eyes are, the more drawn we are to them as a potential mate.

Symmetry isn't the only proportion that signals genetic fitness. Women with waist-to-hip ratios that are correlated with increased fertility (often referred to for obvious reasons as "childbearing hips") have also been consistently rated as more attractive. Though estimates of the magic ratio vary slightly, most scientists agree it's about .70.[8]

Facial features signaling elevated hormone levels (which are also linked to health and fertility) are also generally interpreted as more attractive. For example, in men, those George Clooney–esque features such as a defined jaw and dominant eyebrow ridges that women find so irresistible are correlated with elevated testosterone levels.[9] And in women, elevated estrogen levels are associated with such envious features as high cheekbones and an immaculate complexion (think Audrey Tatou). Even though we may not be consciously registering these signals, they can trigger such powerful urges to be with this person in the short term (that is, lust) that we can quickly forget about the long-term benefits of staying faithful to our less "robust" partner. The mind is loaded with mechanisms meant to ensure that our genes are passed on, and the urge to have sex with an attractive (and thus genetically fit) stranger is one of them.

One of the best demonstrations of exactly how our inner grasshopper tempts us to cheat comes from a very clever study by Randy Thornhill and Steve Gangestad. The experiment sprang from the following theory: our preferences for people with physical cues signaling good genes should be strongest at the moment when we have the most to gain from those genes—that is, when the chances are highest that the interaction will actually result in reproduction. The researchers figured that if the point of having sex with an evolutionarily fit mate is ultimately to pass on one's genes, wouldn't a woman's desire to hook up with an attractive partner be highest when she was most fertile, namely, around the time of ovulation? So they theorized that when women are ovulating—when the benefits of a tryst are greatest in the biological sense—they'd be most attracted to symmetrical features and other physical markers of high testosterone. Sure, they may love their faithful asymmetrical husbands who have invested much in the family, but remember, the balance of power between the grasshopper and the ant is always shifting. So, Thornhill and Gangestad reasoned, at the time when conception is most likely, the grasshopper would be ramping up its efforts to push those women into the arms of a partner—perhaps the guy at the office with the square jaw—who just might have better genetic offerings.

Okay, so how to prove this theory? Here's where things got downright dirty—in a literal sense. The researchers brought women (some ovulating, some not) into the lab and asked them to smell a number of men's unwashed T-shirts and indicate which man's scent they preferred. Keep in mind they never *saw* these men; they simply sniffed their laundry. We know, it sounds a bit strange, but believe it or not, the ovulating women overwhelmingly preferred the smell of men who had more symmetrical features. They sniffed out the scent of genetic fitness, so to speak.[10] In other words, women in the most fertile phase of their cycle preferred the scent of men with whom they probably had more to gain, genetically speaking, by sneaking off into the laboratory closet. This interesting revelation that smell can trigger physical attraction has not been lost on the perfume industry, we might add. One boutique called Booty Parlor, for example, advertises a perfume called Flirty Little Secret, which it claims contains a powerful human pheromone that will make you irresistible to the opposite sex. So those who don't have the scent of symmetry need not fret—there's a way to buy it!

It would seem we should never underestimate a fertile woman's drive to bed a partner with good genes. Work by Martie Haselton at UCLA has documented that around the time of ovulation, not only can women sniff out the "best" mates, they even alter their behavior (sometimes consciously, sometimes not) in an attempt to lure them. How? Well, you've probably never considered how your sense of style might be determined by your fertility, but it turns out that when ovulating, even relatively demure women will dress more sexily—shorter skirts, plunging necklines—when the opportunity for a sexual liaison with an attractive male presents itself.[11] While such behavior—wearing skimpy clothes, flirting, and the like—might easily be dismissed as unbecoming, when we look at it in this context it suddenly seems understandable.

At this point you might be thinking, "Sure, these are all interesting studies, but how does this explain why I woke up in the bed of a stranger when I should have been at home cooking eggs and bacon with Mr. or Ms. Tried-and-True? Because the stranger had a symmetrical face? Because he or she smelled good? I don't think so."

Certainly we don't consciously scan the people at the bar to compare the relative positioning of a potential mate's eyes and ears. Nor do we give each candidate a good long sniff to determine whether he or she is worthy of our fleeting affections. No, our minds do this work for us. Our intuitive mechanisms are so highly attuned to the subtle cues in our social and physical environments that they can direct our attention in a crowded room, if even for the briefest glance, and tip the scales that determine whether and with whom we may try to score, and at what cost.

So we're all equally vulnerable to the temptations of lust, and when we do succumb to them (even if we know the right path to follow is the one that leads to lasting love), it isn't always a mark of a flawed character. Remember, when the grasshopper and the ant go to battle, it isn't always on equal footing. Sometimes we simply *can't* take our mind off someone even though we know we should. Work by the psychologist Jon Maner has demonstrated just how powerful our short-term impulses can be in diverting our attention and desires toward particular individuals. In one of Maner's studies, participants were asked to think about times in their lives when they had been most sexually aroused. Next the researchers asked them to group objects that appeared on their computer screens into categories. Naturally, though, the researchers weren't actually interested in how they categorized the objects. They were interested in how sexual arousal would affect physical attraction.

In addition to the objects, pictures of people who varied in physical attractiveness were also going to appear on the screen, and unbeknownst to the participants, the researchers were really measuring how long it took participants to pry their gaze away from the most attractive faces in order to categorize the objects. If we are programmed to be more captivated by attractive people when the chance for reproduction is highest, the researchers reasoned, the participants would linger longer on the pictures of the attractive faces in the lab, particularly when they were primed to think about sex. Indeed, as predicted, the people who had thought about the sexually charged events had a harder time drawing their gaze away from physically attractive individuals—even at the expense of finishing the categorization task they were ostensibly there to complete. Why? Because their inner grasshopper was so ramped up by thoughts about sex, it overpowered the voice of the ant telling the people to work hard on the task at hand. "Who cares about your job," it seemed to say, "when there are hot people to check out?"[12]

What happens when Mr. Right is wrong

Up until now, we've only been talking about situations where our gut impulses steer us in a useful direction, evolutionarily speaking—in other words, toward mates who will increase the likelihood of us passing on our genes. But what about when those impulses, which aren't always attuned to the world in which we live today, steer us wrong? As most of us well know, the type of person we end up with after a night out isn't always the type we feel good about waking up next to in the morning. This makes a lot more sense when we recall that, as we said in the introduction, our immediate, spontaneous reactions aren't always wrong, but they aren't always right either. You shouldn't consistently follow your gut, nor should you always discount it. Sometimes the physical cues we've been talking about—the structure of a person's face, or the scent of someone's hair—will steer us toward a partner who might provide some short-term enjoyment yet prove to be big trouble in the long run. This is because our emotional impulses have evolved over time, and instincts that may have been beneficial to us thousands of years ago, in environments quite different from the ones in which we live today, are not necessarily adaptive for us now.

For example, while high levels of estrogen may have been the most reliable marker of reproductive success in ancestral environments, with all the medical advancements of the past few decades (fertility treatments, genetic testing, etc.) its predictive value is lessened today. Similarly, while choosing a man whose size and strength could protect offspring from outside threats might have been important thousands of years ago, today there are things more crucial to human survival than being the strongest one on the block. What if, after flexing those bulging biceps, that same man had gone on to spout racial slurs, an ambivalence toward education, and an aversion to the workforce? We're guessing that warm glow you felt the night before will begin to dissipate, fast. When this happens, what's actually going on is that the balance of power between the grasshopper and the ant is shifting, and the voice looking out for your long-term interests (i.e., to have a stable mate who will be a good father to your children)—the one you probably ignored when you decided to go home with this loser the night before—is suddenly getting louder. So while your impulsive decision to go home with this strapping young lad was originally driven by short-term temptations, it's the voice of the ant correcting your initial error and telling you to get out of there and never talk to him again. Ultimately, which voice wins often has less to do with our "character" and more to do with the specific situation at hand.

Take me home for life

When asked to reflect back on their marriage of eighty-two years—believed to be the longest in recorded history—John Rocchio spoke of Amelia's "fine legs," and Amelia Rocchio reminisced about her beau's "handsome presence."[13] But while immediate attraction certainly played a role in the early stages of their relationship, could it possibly explain their sustained affection and fidelity eighty-two years later? After all, even the most irresistible of features succumb to the forces of age and gravity after eight decades. It can't be denied that even if physical attraction was what initially brought John and Amelia together, it was love, not lust, that kept them together all those years. But let's go back to the question we raised earlier: what exactly do we mean by love? As we mentioned at the beginning of the chapter, most people seem content to speak of love in vague and ambiguous terms like "You'll know it when you feel it." But surely we can come up with a better definition for a psychological state that so powerfully impacts so many of the decisions we make in our social lives. For that very reason, psychologists have recently begun to explore exactly what it is that we feel when we say we're in love, and how it shapes such important choices such as whom we pick as a lifelong mate, why we sometimes choose to break our vows, and how we respond when a relationship is suddenly threatened.

Looking for love

So far we've talked mostly about lust—those physical cues that so powerfully draw us to another person. But believe it or not, not *all* attraction is physical. While our grasshopper is telling us to try to score with that symmetrical, fertile-looking beauty or stud ordering the double espresso, our ant is simultaneously trying to divert our attention to the more average-looking character in the corner with the tender and caring smile, or remind us of our beautiful-in-a-time-weathered-kind-of-way spouse waiting at home. After all, short-term sexual conquests are not, especially in these days of birth control, guarantees of reproductive success. Often it's the contrary; in the long run, those who form close and stable bonds with partners make better mates. Which is why our intuitive systems have been equipped with the ability to sense who will be devoted and loyal partners and parents, able to resist the urge to stray—those who will meet the demands of family life, raise healthy and happy children, and protect their families from outside threats.

In other words, romantic love actually helps us solve an important evolutionary problem. How do you know your partner will remain committed to you and your children (and you will remain committed to her or him) in the face of constant temptation? How do you ensure he or she won't run off with the sexy tennis pro, leaving the kids vulnerable and unprovided for? Love, for lack of a better phrase, is the answer. It's the trick our inner ant has up its sleeve to keep us from being cheaters and child abandoners.

Just like its myopic cousin, the grasshopper, the ant too has a hormonal cocktail to steer you down its path. Several studies have found a reliable link between a man's level of testosterone and mating effort; the higher the testosterone, the more effort expended not only in finding a mate but also in competing with rivals for her affection. On the flip side, studies have also found that once a man is in a committed relationship, *lower* testosterone is associated with monogamy. In one, Matthew McIntyre and his colleagues measured the testosterone levels of men in committed relationships and then had them report their interest in having sex with other women.[14] As it turned out, those with higher testosterone levels reported having more interest in playing the field, while those with relatively lower levels were more comfortable with commitment. In short, levels of testosterone can rise and fall depending on whether a man is looking for a one-night stand or is in it for the long haul. And because, as we've noted, women are so adept at (subconsciously) picking up subtle cues that signal high testosterone, this can be a good marker of whether that guy across the room is Mr. Right or Mr. Right Now.

Similarly, women's relative interest in one-night stands vs. long-term relationships is also affected by hormone levels; as we've seen, it varies across the ovulatory cycle. When women are fertile, they report greater feelings of attraction to men other than their partners, especially when these other men have "manlier" characteristics. Yet, by the same token, during less fertile periods, they show more interest in the stability of the relationships they have. In other words, when they're not likely to get a baby out of it, the emotional urges to cheat are quieter for women.[15] And so the balance shifts with the terrain, as the struggle between the grasshopper and the ant rages on within us.

Still, this all leads to the question of how exactly love works to tip the balance toward our long-term interests. Well, it's actually quite simple. After all, as the story of John and Amelia Rocchio suggests, love is the emotional state that binds us to another individual, blinds us with devotion, and compels us to put our partner's interests and well-being before our own. If this is the individual to whom we plan to devote our primary resources, the benefits are obvious. Love, in short, increases our genetic fitness by increasing our desire to be around that other person, thus increasing the opportunities for mating. Countless studies have shown that love is the ingredient that keeps us loyal and committed to our mate.

In one such study, researchers at the University of California, Berkeley, had couples perform a task that was designed to reveal their level of love and affection. They next asked both members of the couple to (privately) report the most significant problem associated with their relationship, and then, in what must have been an incredibly awkward series of interactions, they had couples engage in what inevitably became a pretty heated ten-minute discussion about that very topic. Turns out that couples who had previously been rated as more loving had far more constructive conversations about the contentious topics. In other words, the more loving they were, the more their long-term focus on maintaining a stable relationship trumped their short-term desire to win the argument.

Of course, this isn't terribly shocking. We might expect loving couples to be better at resolving conflicts than unloving couples are. But what was most interesting about this study was how these researchers defined love. Rather than just asking the couples how they felt about each other (which isn't very reliable—after all, how many people are going to openly admit to not loving the person they're with?), they looked for subtler signals. And what they found was that just as there are physical cues that signal sexual attraction, there are also physical cues that correspond with long-term compatibility and affection. This isn't just the speculation of singer-songwriters and romance novelists; in recent years, science has made some very real discoveries about what piques our long-term interest in another person. For example, in experiments where couples are asked to recall and discuss romantic moments

Keeping tabs on your other half: The dual nature of the green-eyed monster

Lisa Nowak grew up in Rockville, Maryland. She was the co-valedictorian of her high school class and graduated from the U.S. Naval Academy with a master's degree in aeronautical engineering before going on to become a captain in the navy and a pilot. Then, after extensive physical and psychological screening, Lisa became one of the 0.7 percent of applicants chosen to be a NASA astronaut—an elite group made up of the best of the best—and in July 2006 joined the crew of the space shuttle *Discovery.* While her career was flourishing, so was her family life; she and her loving husband had just welcomed their third beautiful child. In short, Lisa Nowak was a successful, accomplished, and rational woman by any standard. Why, then, on February 5, 2007, did she drive the nine hundred miles from Texas to Florida wearing a wig, trench coat, and adult diaper (so that she wouldn't have to delay her trip by pulling into a rest stop) while wielding a steel mallet, a four-inch Buck knife, and a loaded BB gun?

Nowak had met fellow astronaut William Oefelein in 1996. They became friends over the course of training and in 2004 began a romantic affair. But after two blissful years, according to colleagues who knew the couple, things began to go sour. And in January 2007, when William decided that air force captain Colleen Shipman was in fact the woman for him, he reportedly broke up with Nowak. By all accounts, this event is what set Nowak off on a *Fatal Attraction*–esque rampage that culminated in her cornering and threatening Shipman in an Orlando airport parking garage.

Everyone who knew Nowak, including the psychologists at NASA who had evaluated her, were stunned to learn that she was capable of such erratic, impulsive, and aggressive behavior. One classmate, Matt Schatzle, said, "She was the closest thing you could get to being a rock star. . . . It seems out of character." Another, Brian Cassie: "She has been an incredible role model as a Naval Officer, astronaut and mother, and has shared her success with many others." NASA staff, scratching their heads, stated that there was "no indication of concern with Lisa. We were all taken by surprise."[17] Her own family even said, "These alleged events are completely out of character and have come as a tremendous shock to our family."[18]

So what happened here? By all accounts, Lisa seemed a reasonable, levelheaded woman. She had been put through some of the most rigorous psychological analysis and testing and had passed with flying colors. After all, she worked for NASA—any hint of mental instability and she would've been kicked out immediately. Yet *something* drove this upstanding citizen, accomplished navy pilot, and mother of three to not only carry on a two-year affair but go off the deep end as soon as another woman poached her lover. What could possibly have possessed her to take that ill-fated fourteen-hour drive that would eventually destroy her family and end her career? What could drive a person of such unassailable discipline and seemingly virtuous character to act so badly? It was jealousy, plain and simple.

Until now, we've been talking about what choices we make in relationships—important choices, such as to whom to commit and for how long. We've talked about the constant psychological battle between love and lust, and why certain forces steer us to pursue short-term sexual encounters while others keep us ever loyal to "the one." But thus far we haven't addressed one of the most powerful tools our inner ant—the side fighting to keep us in a stable long-term relationship with a suitable partner—has at its disposal.

Enter jealousy, the dark side of love. Jealousy is the trump card impulses, so we have different responses to them. We all know people who, like Lisa Nowak, seem completely grounded and normal yet can fly into a jealous rage at a moment's notice. But how common is it for jealousy to produce an aggressive, even violent reaction? How easily can jealousy turn seemingly average, rational folks into vengeful punishers? We hear people say things like "She doesn't have a jealous bone in her body" because most tend to believe jealousy is a fixed trait—that you're either a jealous person or you're not. But as you will probably guess by now, this just isn't accurate.

If our view of character is correct, then this desire to guard what's ours lurks in all of us. It just needs the right conditions to emerge. But what are these conditions, exactly? To answer this question, we headed back to the lab. We had one big problem, though. In designing an experiment to study jealousy, we couldn't exactly go around trying to trick people in existing relationships into cheating. It's not only ethically questionable, it's downright dangerous! Case in point: When Dave was a graduate student at Yale, one of the favorite stories told every year involved a study on jealousy that had used real couples. The experimenter had brought the unsuspecting couples to the lab and seated them in separate rooms. He then told the woman that he was going to flirt with her as part of the experiment and that she should play along. After turning on the intercom system so that her boyfriend could hear everything that was happening, he propositioned her, and waited to see how the boyfriend would react. He didn't have to wait long. Unfortunately for the experimenter, the boyfriend in one of the first sets of couples just happened to be a very hotheaded member of the Yale football team. He immediately burst into the room, and before the experimenter could explain, the linebacker landed a punch. Thus ended that particular experiment.

Anyhow, when we set out to study jealousy, we decided we

wanted to keep all our teeth (plus the ethics review board at our university didn't think it was such a great idea to make people believe that their partners might be open to the advances of others). So instead of messing with existing relationships, we did the next best thing—we created some new ones. We brought single people into a lab and had them begin to form a rapport with a flirty partner, only to have it dashed by the arrival of an alluring rival (who of course was working for us). What's more, to see how easily jealousy triggers revenge seeking, we also gave subjects an opportunity to punish their fickle love interests, as well as the man- or woman-stealers.

The green-eyed monster in all of us

Imagine you're a female freshman in an introductory psychology course. Your professor has informed you that as part of the course requirements you will need to take part in several experiments being conducted by researchers at your institution. You're not particularly interested in participating, so you sign up for the experiment that seems to require the least level of time and effort, the one called "Group Problem Solving." After all, if you're doing something in a group, that's less work for you, right? You show up grudgingly at the scheduled time, thinking about all the better things you could be doing.

As you settle into your seat, in walks another participant of the opposite sex who you are told will be your partner. He sits down in the chair next to you. Once the experimenter leaves the room, he turns to you and starts to chat. It's just small talk at first, but after a bit you two are joking and chuckling together. He's moved his chair closer and is leaning toward you, making eye contact, and sending you strong, unmistakable signals—this guy is full-on flirting with you. Suddenly you think: "Hey, he's not offensively unattractive. Maybe something good will come of this experience after all." Unfortunately, the experimenter comes in to begin the experiment and interrupts the moment. But after you're given an introduction and asked to perform some preliminary tasks individually, it's time for the "group problem solving" with the attractive and charming partner. Now the flirting is back on, in full effect. His chair is closer than it was before, and you two are swapping stories and laughing and having a grand time. He is definitely interested in you; you're sure of it.

Suddenly your reverie is interrupted again by a knock at the door. Another participant has arrived, ten minutes late, and she's pleading with the experimenter to let her take part in the study so she won't have to come back another time. You and your new love interest roll your eyes at each other, but the experimenter relents and escorts this interloper into the lab. Now three of you are working on the group task together, and the guy scoots over, away from you, to make room for the new woman. Things go smoothly at first. But after a while you notice that the other two are starting to generate a rapport. Now *they're* laughing at each other's jokes. He's starting to lean more toward her than you, and it seems like he's paying less and less attention to you.

Now the experimenter comes back into the room wearing a concerned look and explains to the three of you that there's been a mistake. It turns out that the woman never should have been allowed to join the group; this particular task was supposed to be completed in pairs or alone. So he states that the group of three is going to have to split up. Without hesitation, the guy turns to the other girl and says, "Hey, why don't we work together?" She relationship to our sense of self-worth. Mark Leary, a psychologist at Duke, has argued persuasively that self-esteem acts as a sort of social barometer (a sociometer, if you will) that goes up when others like us and down when they don't.[20] One of its functions, then, is to motivate us to protect our relationships. After all, if having someone like us enough to be in a relationship makes us feel good about ourselves and having those relationships threatened makes us feel bad about ourselves, we're obviously more invested in making that relationship work out. It's human nature; we're designed to care about how others view us, and to allow them to help define who we are. Think about how it feels when someone you like or love doesn't seem to like you back. Painful, right? And recall Plato's story of the origin of love. While he may have gotten the details wrong, there is some psychological truth to the idea that the one we love can come to feel like a part of us. Once this bond has been established, jealousy can be the glue that protects that whole from splintering again.

To see if our theory about the link between jealousy and self-esteem was correct, we turned to these same subjects, asking them a series of questions designed to reveal how jealous they felt after the interaction (though we had a good idea of who would be jealous, since we made them feel that way on purpose), followed by another task that tapped into implicit feelings of self-esteem and self-worth (a variation of a task known as the Implicit Association Test, which is designed to get at people's associations or opinions without directly asking them). Indeed, it turned out that the participants who felt jealous also demonstrated lower feelings of self-worth than did those in the control condition, who, instead of being snubbed, were told Carlo had to stop working with them because he "just remembered" a long-scheduled medical appointment.

What was most interesting to us about these results, however, was that such a small slight could so drastically affect people's sense of self and emotional state. If you look at the author photo on the jacket of this book, you might be skeptical—*that* guy made people jealous? Just by choosing to work with another woman on a social psychology experiment? Yeah, that's right. And not just a little jealous either. Some participants' faces quite literally dropped when Carlo threw them over for the other woman. Others let out audible gasps. One participant found the rejection so unbearable, she repeatedly shushed Carlo and the rival as they worked on their tasks, angrily sneering, "I can still hear you" when they continued to joke and giggle together on the other side of the room divider. All of which speaks to the power of jealousy and how quickly it can rear its ugly head to protect even the *potential* for a relationship.

But there was still more to this experiment. It was now clear to us that jealousy is very easy to instigate and that it emerges when our self-esteem is threatened. However, there's a big difference between just feeling jealous and actually going so far as to act on it aggressively. Consider the sad case of Stefanie Rengel, a fourteen-year-old Toronto native who was stabbed to death outside her home on New Year's Day, 2008, by her former boyfriend, seventeen-year-old David Bagshaw.[21] But David wasn't even acting out of his own jealousy. He was acting at the behest of his current girlfriend, Melissa Todorovic. Allegedly Melissa had become so consumed with the idea that David still had feelings for Stefanie that she ordered him to kill her. In a series of text messages, she wrote: "I want her dead . . . lol we've been through this . . . If it takes more than a week then we're just going to be friends." Melissa had never even met the girl on whom she had called for a hit. The mere idea that Stefanie might be stealing David's affections was enough to throw Melissa into a murderous jealous rage.

was appropriate, and that the other two (the snubber and the rival) would have to consume every last drop.

Then we left the room and gave the participant some time to read the questionnaires (who could resist?) and prepare the samples. And when she read the questionnaires, what did she see? That these other two hated spicy food (naturally, we'd rigged the questionnaires to read this way), which was what she'd been randomly assigned to give them. So now she was faced with a difficult decision: how much extra-strong hot sauce did she want to give this sleazy, no-good jerk and his floozy? Turns out it wasn't that difficult a decision after all. The jealous participants loaded up the sample cups with that painful stuff, filling them with significantly more than those in the control condition did. As it turned out, how much hot sauce they poured was directly predicted by how jealous they felt.

Now, making someone who hates spicy food ingest large quantities of the world's hottest hot sauce may not be quite the same as going after your ex or his mistress with a steel mallet and a Buck knife, but it stems from the same underlying desire: to punish. In the lab, we can't create a relationship of many years, but every relationship has to start somewhere, and if we don't have mechanisms looking out for what *may* develop into a long-term relationship, then we may never keep that long-term partner around in the first place. As this experiment revealed, this mechanism is jealousy. After all, all these participants who became jealous enough to want to punish their partners and rivals with hot sauce were ordinary, run-of-the-mill college students.[22] Most of them wouldn't have thought they would feel jealous in such a situation, and definitely that they wouldn't leap to get revenge. But leap they did. And it's all because their inner ant was pushing them to protect what *might* have turned into a happy, stable long-term relationship. When you think about it, this urge to hurt those who would steal our partners, and even sometimes our partners themselves, in order to keep them with us or make them come back to us makes some sense evolutionarily speaking. While such a strategy may not have helped our ancestors keep their partners happy, it certainly would have helped to keep them in line

It's true that many people think of jealousy as a character flaw. But if we didn't feel jealous, we wouldn't have the kinds of stable relationships that are necessary to adequately protect and care for our offspring. It may not be a pleasant emotion, but sometimes it can be a quite useful one, at least when experienced in mild doses. It can alert us to signs that our partner is being unfaithful or that someone is trying to steal him or her from us. It can also signal to our partners that we want to be in the relationship for the long term (otherwise it wouldn't be worth putting up a fight), and signal to us when they feel the same.

Obviously, when experienced in too great a degree, jealousy can get out of hand. Whether our preferred weapon is hot sauce or BB guns, we all have the potential, as we've just seen, to succumb to its darker side. The upshot, then, is not always to ignore feelings of jealousy or write them off as a fault, but rather to try to understand how to make them work for you. How? Pay attention to jealousy when it first arises; allow yourself to experience it and don't shut down your emotional intuitions. Ask yourself what's causing it and why. Think hard about whether the jealousy you're feeling is irrational or warranted, then use that information to fix your relationship before it's too late. If you don't allow yourself to experience jealousy in the first place and never learn to analyze the resulting feelings, you're leading yourself down a long, dark, and destructive road, because at some point the pressure behind those feelings will explode, and jealousy will take charge in a way that is out of your control—just like it did for Lisa Nowak. Remember, the emotional impulses that were adaptive thousands of years ago may be useful in different ways today. Sure, acting on every jealous urge may have made sense long ago, but in the modern world there are laws against stalking your partner and beating your rival with a club.

The fine line between Casanova and Ward Cleaver

With love and jealousy at its disposal, your ant has quite an array of tools to guide you toward forming and keeping lasting relationships. Of course, these forces are constantly competing with the weapons on the grasshopper's side. It's an ongoing tug-of-war to determine whom we bed and whom we wed. Thanks to all the tricks the mind plays on us, predicting whether love or lust will ultimately win out can be complicated. As anyone who has ever been in a serious relationship knows, what we think we want in a partner isn't always what's best for us. While we have certain automatic impulses and mechanisms that have evolved to help us meet our romantic goals, they can be co-opted by context or overwhelmed by the forces working for the other side.

So which side to root for? There is no easy answer. Both can get you in trouble, and both can give you exactly what you want. Both can make you act in "good" and "bad" ways, and both are sensitive to information you are not consciously aware you have. How do you make the right decision? How do you know when to give yourself over to lust and when to hold out for true love?

Whether to channel Casanova or Ward Cleaver? The first step is to understand your goals. Are you looking for Mr. Right, or will you be happy with Mr. Right Now? Are you trying to stay faithful to the one waiting for you at home, or are you trying to find someone new? When you're in the moment, choosing lust over love is relatively easy. That's not rocket science. But what about when you know you should choose love? How can you help yourself resist the powerful temptations of lust? Well, the first step is to avoid situations where these temptations may arise. Sounds obvious, but think about how easy it is to succumb to temptations after five tequila shots.

Odysseus, the main character of Homer's epic *The Odyssey*, seemed to have this problem figured out. On his long voyage from Troy back home to Ithaca, where his wife, Penelope, was awaiting his return, Odysseus and his men passed the island inhabited by the Sirens, the irresistible seductresses who used their mellifluous voices to lure men to death by shipwreck against the rocky shore. Odysseus understood that he and his men would be unable to resist the Sirens' song, so he had the crew fill their ears with wax and, as an extra precaution, had his men tie him to the mast of the ship so he'd be unable to steer them toward a rocky doom. It seems the Greeks knew something about character: that we can't count on being able to control our temptations by sheer logic or will.

So if your goal is smooth sailing in your current relationship, avoiding temptations might be the path to travel. Do a bit of soul-searching. Knowing your weaknesses—those cues that make you forget about your soul mate and unleash your inner playmate—will help you know when to tie yourself to the mast and when it's safe to go full steam ahead toward shore.

Of course, how a relationship turns out isn't always up to you.

As they say, it takes two to tango. Problems can also arise when you want one thing and your partner wants another. When one person is looking for a fling and the other wants a life partner, all bets are off. This is when the green-eyed monster often takes over. As we've seen, the potential for jealousy resides in each of us, no matter how levelheaded we think we are. If participants in our experiment felt jealous after only flirting for a few minutes, and if they were willing to cause pain as a result of that jealousy, imagine what can happen when the relationships are more established and meaningful. Very few of us will probably don a wig and a diaper to threaten our rivals at gunpoint, but most of us, whether we want to admit it or not, would probably take the opportunity to make life miserable for our partners and those who poached them. Our character can transform from caring partner to vindictive ex with ease.

4 / From Pride to Hubris

The deadliest of the seven sins?

Thomas Mapother had something of a difficult childhood. Ever since he was a young child growing up in Syracuse, New York, he and his family seemed on the move. His father worked as an engineer for General Electric, a job that required him to relocate the family quite often. Thomas and his three sisters must have felt something like nomads as they moved to Ottawa, Missouri, New Jersey, and then Kentucky. Such moves would be hard on any child, but they were all the harder on the Mapother family, as Thomas' father was not the kindest or most trustworthy of men. In fact, as Thomas would later recount, his father was so unpredictable and abusive that the kids' mother, Mary Lee, finally divorced him, taking the kids with her.

The months and weeks following the divorce were quite difficult both psychologically and financially—the family was living near the poverty line. Although the Mapothers accepted food stamps to help with meals, they refused to take welfare. They were a proud family and weren't about to take handouts if they could help it. So everyone went to work. Thomas in particular felt the burden, since even though he was young, he now felt he had to take on the role and responsibilities of the "man of the family." He got a newspaper route and put his weekly earnings straight into the family coffer. For better or for worse, Mary Lee married again relatively quickly—this time to a plastics salesman named Jack South—and the family was on the move yet again.

All the moving around was starting to take a toll. Thomas had attended fifteen different schools by the age of fourteen, and had been bullied in a majority of them. He was always the new kid, and as such, he always had something to prove. With each new school, he had to show his often unwelcoming peers that he was tough enough, and his always-demanding teachers that he was smart enough. To make matters worse, Thomas was quiet by nature and also suffered from dyslexia. He wasn't the kind of kid who became immediately popular, nor the kind who was instantly flagged as the class brain. Yet he felt a deep sense of pride that drove him to work relentlessly, pushing himself harder and harder in an attempt to prove himself to others. He toiled for hours at night trying to make sense of the pages of his schoolbooks, a task made even more difficult by his dyslexia. When his family finally settled in Glen Ridge, New Jersey, for his last few years of high school, he joined a number of sports teams. He wasn't a gifted athlete, but he still managed to impress his coaches with his intensity, if not his ability. Thomas had pride in himself and in everything that he did, and it was infectious. It was hard not to root for him.

Thomas quickly was dealt a blow, however, when a knee injury from wrestling put an end to his athletic career. But as one door closed, another opened. He tried out for the school production of *Guys and Dolls* and was soon bitten by the acting bug. With sports now out of the question, he threw himself into acting with the same single-minded resolve that marked all his endeavors. He took great pride in his growing thespian skills and quickly set his mind to making it as an actor. At eighteen he moved across the river to New York City, where he began pursuing his new career in earnest. By day he took any job he could find—waiter, busboy, porter—and by night he took acting classes. He auditioned for anything and everything he could find. *He* knew he could be the best, but he realized that the only way he was going to reach the top was through a single-minded focus and the highest devotion to his craft.

After many arduous years, this perseverance finally paid off. Thomas Cruise Mapother, who by now had shortened his name to Tom Cruise, was offered an audition for a one-line part in the movie *Taps*.[1] He approached the audition with such intensity and such confidence that Harold Becker, the director, decided to give him a higher-profile role, with billing to match. From that point on, Tom seemed to be a golden boy. He won role after role, and soon became one of the most popular actors in America, with a résumé and earning power that were the envy of his peers. A 1989 issue of *Time* magazine proclaimed, "With each adventure, audiences adjusted their estimation of the young man—from Most Likely to Succeed to All-American Dreamboat to Serious Actor worth taking seriously." As Jeanne Tripplehorn, his costar in *The Firm*, put it: "He's absolutely gotten better through the years—and seems to be evolving into a man of great character."[2] Yet all the praise didn't seem to be going to his head. He was still just as dedicated to his craft, such a picture of pride and professionalism in his work that as early roles. Without it, we probably never would have heard of Tom Cruise. Yet at a certain point in his career, pride turned suddenly to hubris, a hubris that was on very public display as he lectured Matt Lauer on his moral superiority, and as he "jumped the couch" on *Oprah*.

So what pushed him over that treacherous line between pride and hubris? Despite what some might think, Tom's seeming shift in character had nothing to do with his adopting the doctrines of Scientology. No, Scientology, with its message to the elite, was just a symptom, not the cause. When you think about it, it's a familiar story. Whether it's titans of Wall Street such as Richard Fuld, who rose from being a hardworking trader at Lehman Brothers to the arrogant CEO who was largely responsible for the company's collapse, or politicians such as former senator John Edwards, who rose from humble beginnings in a North Carolina mill town and ended up as the poster boy for self-indulgence, we've all seen how easily pride can turn into hubris when the temptations are ripe. What may not be so easy to accept, however, is that like the other character "flaws" we've been talking about, this can happen to anyone. How, when, and why pride turns into hubris is a complicated question. And as you'll find, it's one that has surprising answers.

Pride's PR problem

Pride—it's got a bad rap. Those of you who spent any time listening to sermons (like we did) may remember that not only is pride one of the seven deadly sins, it's considered the deadliest of them all. And that's not just in Christian theology either. In Buddhist teachings pride is one of the ten fetters that prevent enlightenment, and the story is much the same in the Torah and the Tao Te Ching. Even modern-day secular texts paint pride as a moral failing. More than 90 percent of the synonyms for *pride* in any English thesaurus have a negative connotation: *arrogance*, *conceitedness*, *being full of oneself*, and the like.

Can this be right? Sure, sometimes pride can get out of control and land you in dangerous waters, but if pride is *always* to be avoided, some facts just don't fit. Yes, pride can lead to poor decisions if we assume that we can do no wrong; yet can't taking pride in our work actually motivate us to do a good job? Yes, we hate hubris in a leader, but (if the past few presidential elections are any indication), we also don't take to following timid wimps. The way to make sense of these seeming contradictions is to realize that pride may not be what you think it is. Like all the supposed character traits we're discussing in this book, pride isn't a fixed feature of our personality, nor is it always a vice. By the same token, humility isn't always a virtue. What may be even more surprising is that pride doesn't always come from within; it often emerges as a response to what's going on around us. So whether you exhibit pride or hubris, whether it helps you to lead or makes you despised, is all determined moment by moment as a result of the ongoing battle in your mind between the ant and the grasshopper.

Pride and perseverance: Working hard or hardly working?

What motivates us to work hard? To persevere, like the young Tom Cruise, in the face of obstacles and challenges? One answer, we believe, often boils down to pride. If you were thinking the answer is a deep-seated desire to succeed or to be recognized, you're not entirely wrong. These desires are indeed linked to pride. But contrary to what you might think, pride doesn't always come from satisfying some goal we set for ourselves. No, pride, at least to start, stems from external factors in our social world. Put differently, what makes us feel proud isn't always up to us.

One of the biggest challenges we face in life is the universal quest for social status. Status is something we all want. After all, the higher you climb on the social ladder, the more desirable you are as a leader, friend, colleague, or spouse. In terms of long-term rewards, high status means you're golden. Yet making your way up the ladder usually requires a lot of hard work: developing skills as an athlete, getting top grades as a student, earning promotions at work, building a huge network of friends. Sure, it might be easier to sit at home and be a couch potato, but this option, although perhaps pleasant in the short term, yields few rewards in the long run. Gaining status, then, is a classic battle between our dueling interests. While our inner grasshopper just wants short-term pleasure, the ant knows that in the long term, when the playing is done, people look to, rely on, and reward the experts, the leaders, the ones who got ahead. But, of course, the way to be one of these admired few and to reap the associated benefits almost always involves some level of perseverance. That's where pride comes in.

If our view of how pride works is correct, two things must follow. First, pride has to be socially determined to some degree. After all, if the whole point of pride is to make us valuable to the group, then the qualities that makes us proud would have to be something the group deems important. If this is the case, we should be able to make people feel proud of anything—even trivial things they've never cared about before—just by giving them a sense that these things are valued by peers. Second, this feeling of pride should motivate people to work harder and longer, even at things that are very onerous. If both these hypotheses turn out to be true, we'd have pretty convincing evidence that pride is the engine that drives you to cultivate and demonstrate the skills and abilities, whatever they may be, that will raise your long-term social standing.

Sculling to spinning: It's the pride that counts

It was five o'clock on yet another gray and chilly morning in Portland, Oregon, when Lisa Williams was awakened by the shrill sound of her alarm. Though still groggy, she roused herself, and, as she did most mornings, dragged herself out of bed and into the cold morning to meet her sculling teammates on the banks of the Willamette River. By the time she returned to her apartment several hours later, she would have blisters on her hands, her arms and back would be sore, and her clothing would be soaked through. But there was no time for rest: it was on to a full day of grueling classes at Lewis and Clark College.

Why this self-inflicted torture? That's a good question, and one that was on Lisa's mind when she arrived in Boston to join our lab group as a doctoral student. Sure, it was good exercise, but a workout at a more human hour, in the campus gym, with its music, television, and better access to a warm latte, would have been much easier. Yes, being on a college sports team looked good on a résumé, but there were much easier ways to appear well rounded than waking at five o'clock every day to endure physical labor in the damp and cold. So what kept her going? Lisa had a sneaking suspicion that it had something to do with pride.

It was an idea that resonated with our group. The only problem was, how could we explore it experimentally? We realized that if we wanted to test whether pride motivated people to persevere at something, all we had to do was give people a difficult or unpleasant task, make them feel proud of their ability to do it, and then watch to see if they would work longer and harder at it as a result. Seemed simple enough, but there was a catch. The easiest way to induce pride is to tell people that they did well at something. But research by Albert Bandura at Stanford has shown that simply knowing you can do something well can increase your willingness to do it, whether you feel proud or not, as long as it's something that interests you.[5] So if we were going to show not only that pride could be socially determined but also that it leads to perseverance, we had to be able to separate knowing you're competent at something from feeling proud about it. Luckily, Lisa came up with an ingenious idea.

We decided to create the following situation. We'd bring participants into the lab and have them complete a long, difficult, and tedious task. We'd then do one of three things before asking them to work on a second onerous task that they believed tapped the same abilities as the first. We'd either tell them nothing about their performance on the first task, perfunctorily inform them that they obtained a high score on the first test, or give them the same information about their high score but add praise and acclaim. If Lisa's suspicions were right, then only the participants who'd been given acclaim would persevere longer on the second task. That is, simply knowing they're good at a task wouldn't be enough to motivate them to spend more time doing a similar one if it was unpleasant—but pride would.

Here's how it worked. You arrive at our lab believing that you are going to take part in a study on something called visuospatial cognition. As you settle into your seat at one of the computers, Lisa tells you that you'll be completing several tasks designed to gauge how your mind processes spatial relations. If you're like most people, this statement probably translates as follows: *boring*. But it gets worse. "What will happen," Lisa goes on, "is that you'll see arrays of different-colored dots flash on your screen for two seconds at a time. Your job is to estimate how many of those dots are red." Sound like fun yet? After another internal yawn, you turn toward the screen of your PC. Different-colored dots, ranging from ten to forty in number, flash on the screen in various patterns, and after each trial the computer asks you how many of the dots were red. In actuality, there always were too many red dots to count in the time allotted, but not so many that people would feel that there was no way to produce a meaningful guess. Of course, we didn't care how many red dots people saw; our aim was only to give people a task that they believed measured some kind of trivial ability—a task that seemed difficult enough to be annoying, but not so hard that people wouldn't believe us when we told them they'd done well.

Now here's where the social factor came in. Our pilot testing showed that, as we suspected, most people didn't seem to care all that much about what their visuospatial skills were, let alone even know what the term meant. Nonetheless, we expected that if we could lead them to believe that people around them (i.e., Lisa) thought the skill was important, they would all of a sudden feel proud of this new ability they'd never known they had (remember, our theory is that pride, at its base, is a social phenomenon). So if you were in this third group of subjects, after you finished the tedious task of estimating red dots, Lisa would reenter the room brandishing a piece of paper and a very impressed look. As she handed you the paper with your percentile score near the top of the curve, she'd smile and shake her head a little in disbelief while saying, "Good job! That's one of the highest scores we've seen!" Then, after a few questions designed (unbeknownst to you, of course) to gauge how proud you were feeling, she would leave you with another task that again assessed your visuospatial ability. But before she left she would inform you that there were far too many questions for any one person to answer, and so people were only being asked to work until they felt like they had done enough; then they were free to go.

For the other two groups, the interaction with Lisa differed only slightly. In one, Lisa simply handed participants the report that noted their superior percentile rank without commenting on the score. These people would know they performed well, but the ability wouldn't be marked as praiseworthy or socially valued.[6] In the other, no feedback of any kind was given. Lisa just entered the room and started people on the second task.

This final stage consisted of mental rotation problems—the kind where you have to decide if one strange three-dimensional shape can be spun in space to match the image of another. Let's just say that for most people this is about as fun as a root canal. Of course, what we ultimately cared about wasn't how well people did but rather how long they continued working before clicking the quit button. In a nutshell, we wanted to know how long they would persevere.

The results couldn't have been any clearer. Not only did the people who received acclaim suddenly feel pride for a skill they hadn't cared about a few minutes before, but this pride—stemming from the simple fact that Lisa seemed impressed with their score—was all it took to make them work longer and harder on the awful mental rotation task. Those who were made to feel proud of their visuospatial abilities persevered much longer than those who knew they scored well but weren't made to feel proud or those who received no feedback at all. In fact, the more pride they felt, the longer they worked.[7]

What these findings clearly show is that pride and the resulting perseverance didn't have anything to do with the nature of the task or the ability it represented. The task was difficult and tedious and measured a skill that most of the participants never thought about. After all, for most people, being good at visuospatial cognition is just about as important as being good at stapling. Yet simply because we marked this ability as socially important, by having someone care about and give praise for a good performance on it, our participants suddenly felt pride and worked their butts off. Albeit on a smaller scale, we see pride working here as it did for Lisa in her morning sculling workouts and for the young Tom Cruise, giving them the focus, drive, and perseverance needed to reach their goals. In short, we see that pride can lead to long-term success, not just to sin and ruin.

Now, if you're like most people we know, a nagging question may be lurking in your mind. These findings clearly show that what makes you feel proud can be determined to a surprising degree by what others around you seem to value. But what about what *you* value? We all know some people who take pride in things that seem, to put it simply, odd. For example, every year at the Illinois State Fair, there is fierce competition to be crowned the king (or sometimes queen) of the hog-calling contest. That's right, people spend all year suffering sore throats and splitting headaches as they practice getting their squeals to be just the right pitch. In South Cheshire in the United Kingdom, people come from as far away as Australia to compete in the World Worm Charming competition, where they demonstrate their finely honed techniques for stabbing pitchforks in the ground to lure hundreds of worms to the surface. Granted, these may seem like ridiculous or extreme examples, but the point is that people can take pride in idiosyncratic skills or talents that few others care about. On a personal level, both of us know this to be true. We're academics, after all; we have colleagues who take pride in such things as a comprehensive knowledge about the reproductive habits of dung beetles, understanding the differences among the six flavors of quarks, or the ability to read Sumerian texts.

How can we square these very personal variations with our argument that pride is a function of those around us? The answer is simple once you take into account the difference between the older and newer parts of our brains. Remember that the intuitions and emotional responses that seem to emerge spontaneously in us are the results of the older mind's attempts to ensure our survival. Way back when, evolutionarily speaking, the abilities or attributes that were worth developing were the ones that would raise your status and value to the group, as this would help you build a strong social network to rely on to protect you from predators, share resources, and make you a more desirable partner. Laziness wouldn't be a wise long-term strategy, so pride, thanks to the mechanisms of the ant, would push you to get up and work. As the mind evolved, however, the notion of the self became more elaborate. We are one of the few species that have the ability to take a third-person view of ourselves and, thereby, the ability to construct our own social world. That is, we can see ourselves as we are seen by others—we can be our own audience. And as we noted in the introductory chapter, this capacity for abstract thought, which evolved fairly recently in the grand scheme of things, brought about a fundamental shift in how some psychological processes work.

When it comes to pride, what this concept of "self" means is that we don't need someone else to tell us what skills or attributes matter; we can now tell ourselves. This isn't to say that the regard of others stopped mattering, or that pride comes only from within. If this were the case, our experiment never would have worked. After all, who really cares about how good they are at visuospatial cognition? What we're saying is that thanks to the ability to reflect on ourselves, we now experience the pride that comes from within the same way we experience praise from others. We, in essence, can be our own peers. Remember, pride evolved as a way to ensure our social status and, therefore, success in the long run, and it still works this way. But it's important to remember that it's now both internal and external audiences that we care about.

Pride and status: Is he a leader or an ass?

Mission accomplished—at least according to the banner hanging over the deck of the USS *Abraham Lincoln* on May 1, 2003, as George W. Bush became the first sitting president to land a plane on an aircraft carrier. The media coverage was impossible to miss. Here was the president, in full pilot regalia, strutting across the deck to give a speech proclaiming a military victory—the end of hostilities in Iraq. Everyone ate it up. Sure, you might expect gushing from the conservative pundits, but this time partisan ideology didn't seem to matter. Chris Matthews called Bush's appearance an "amazing display of leadership."[8] Keith Olbermann proclaimed, "We're proud of our president. Americans love having a guy as president, a guy who has a little swagger."[9] Even the liberal columnist Joe Klein had to admit, "That was probably the coolest presidential image since Bill Pullman played the jet fighter pilot in the movie *Independence*

Day."[10] On that day, George Bush was the very embodiment of a proud leader of a proud nation, and we loved it.

Of course, it didn't stay that way. As it became clearer and clearer that the Iraq war wasn't near being won, and as other tensions flared in the region, what at first was a symbol of national pride became an albatross, a symbol of the president's hubris. By the time he left office, George Bush had gone from being one of the most popular to one of the most despised presidents in modern history.

Of course, swings in popularity aren't uncommon in politics. In November 2008, many Americans couldn't have been prouder to have elected Barack Obama, an African American man whose name had become synonymous with the call for change. Just a few months into his presidency, he was one of the most admired leaders of recent time, with over 70 percent of Americans reporting they saw him as a strong leader, a likable person, and a source of confidence for the country. Yet by the beginning of 2010, many of those same voters believed they'd made a mistake. They felt that Obama had been overconfident in his ability to bring about change, that he'd talked a big game. Having pushed hard for major changes in health care and environmental policy, he was cast by the opposition as arrogant, and his popularity had plummeted further in less time than that of almost any other new president in history, resulting in Republicans retaking the House of Representatives in the November 2010 elections. Of course, for our purposes, what one feels about either man's politics is irrelevant. The point is that in both cases, and in countless others, the very same quality—pride—that once seemed so virtuous in a leader had become a vice. It had become hubris.

Which brings up another question: Do we like proud people? Do we admire them or despise them? Or, to ask it slightly differently: is acting proud something that signals a "good" character or a "bad" one? We've just spent some time trying to convince you that, contrary to the common view, pride can be a virtue—it motivates you to work hard and persevere so that you can reap long-term social benefits. But if the overall goal is high social status, then shouldn't pride damn well make people like you? After all, isn't that the whole point of navigating social life? We've already established that while in the short run selfishness can sometimes be beneficial, in the long term having friends we can rely on is something we can't live without.

So if pride brings status, why are overly proud people sometimes so detestable? We believe that the trick to untangling whether or not pride is an attractive element of character rests on one critical distinction: whether we're talking about pride that is earned (often termed authentic pride) or pride that is unearned (often termed hubris). Telling the two apart, and even controlling which one you'll exhibit in any given situation, is a complicated business. But before we tackle that question, we first have to prove that pride can breed liking in the first place.

The look of pride

Close your eyes. Imagine something that makes you feel proud—maybe your golf handicap, or your killer homemade lasagna. Notice anything different? Did your posture just get a little straighter? Are you holding your head just a bit higher? Is your chest expanding just a tad? Believe it or not, these are all actual physiological changes that people have been found to exhibit when they're feeling pride. Why? Well, if pride is going to help raise your social status, it probably should have a physical marker, a visual shortcut of some kind, to let others know you have it. Just as Rolexes and BMWs are cultural signals of high status, the mind has imbued the body with certain physical signals of high status, cues that we all subconsciously use to judge whether someone is capable, confident, and leadership material or low on the ladder of importance and power.

Surely you've heard the common expression "puffed up with pride." Well, this is where it comes from. The fact that these physical changes do seem to occur spontaneously, often without us even being aware of them, suggests that they are fairly ancient markers—markers from a time when we didn't have medals, promotions, or Prada. Jessica Tracy and her colleagues at the University of British Columbia have conducted some of the best work documenting how we recognize pride in others. They were the first to scientifically identify the markers noted above and to show both their innateness and their universal link to perceptions of status. In a series of experiments conducted across various cultures, Tracy and her colleagues found that everyone from Canadian college students to members of an isolated West African ethnic group easily recognized and identified the expanded posture and head tilt as being signs of pride. These automatic expressions of pride are so innate and so deeply ingrained, in fact, that Tracy and her colleagues even found them to be exhibited by congenitally blind athletes upon winning events in the Paralympics.[11] These athletes couldn't possibly have learned what pride looked like by watching others, yet they still displayed the very same physical signs.

Now, if people seem to show these signs automatically and universally, it follows that they must have some communicative power, or that others must recognize them for what they are. And indeed, work by Tracy and her group showed that not only do people agree on what pride looks like (and emit the same signals themselves when they are proud), they also intuitively link these physical expressions of pride to status. In another series of experiments, Tracy and her colleagues used a common psychological tool, the Implicit Association Test (IAT), to find out what the mind automatically associates with these physical expressions (as we noted in the previous chapter, psychologists consider measures such as the IAT, which require the brain to make speeded-up categorizations of stimuli, a much more accurate window into people's true beliefs than self-reporting measures such as questionnaires). What Tracy and her colleagues found time and again was that these physical expressions of pride automatically triggered associations with words such as *dominant* or *important*, which clearly indicated perceptions of high social status.[12] In other words, in the blink of an eye, without any other evidence to go on, these subtle physical cues are enough to make people see another person's character in a very specific light.

Now, keep in mind that while this study found pride makes people appear to have high social status, it doesn't say anything about whether it makes them seem more likable. As the victim of any bully or overbearing boss knows, being dominant doesn't necessarily make you attractive. But in the end, isn't that the goal, at least for members of a social species like us? All of which brings us back to the earlier question: are proud people likable? To untangle this question, we went back to the lab to see if people working in real groups and on real problems would follow *and* like proud people, or write them off as arrogant asses.

Follow the leader

"Great," sighed Claire. "I hate group projects." Claire had arrived just a few minutes earlier to take part in an experiment advertised to investigate problem-solving strategies. She was now sitting in a room with two other women, Maya and Ashley. Like Claire, Maya was an undergraduate who'd agreed to participate in our study. Ashley, on the other hand, worked for us, though naturally the other two didn't know it.

Our colleague Lisa, who was again leading the experiment, told the three of them that they'd be completing two spatial problem-solving tasks. One would be completed individually and the other as a team. "Okay," Lisa said. "Turn to your computers and complete the first task. When you finish, come into the next room one at a time for a vision test." The first task was a mental rotation task similar to the one we described earlier in this chapter. Much as before, computer screens repeatedly presented two partially unfolded Rubik's-cube-like shapes. The task was to decide if the one on the left could be rotated in some fashion to match the one on the right. After each person finished, she headed into the adjoining control room of the lab to find Lisa.

Maya was first. "Come in and have a seat," Lisa said, and guided her to a chair in front of a computer. On the screen was a vision test that took a few seconds to complete. "We're just having you complete this to make sure everyone is seeing the images in the same way," Lisa informed her. Then she sent Maya back into the room where the experiment was taking place. Next came Claire. Same test, same explanation from Lisa. But this time, as Lisa was finishing, the printer whirred and out came an official-looking score sheet. "Wow," Lisa said to Claire as she glanced at it, "you scored in the ninety-seventh percentile on that mental rotation test—that's one of the best scores we've seen! Great job!" And back Claire went into the other room, suddenly feeling proud because of the praise that, unbeknownst to her, had nothing to do with how she actually performed. Next came Ashley, who chatted with Lisa for a few minutes before going back into the room to continue the charade (Ashley, after all, wasn't really part of the experiment, so there was no need for her to go through the motions of actually taking the vision test).

Now it was time for the group task. Lisa sat the three women around a small table and showed them a wooden cube made up of twenty-seven smaller cubes attached by hinges. She then unwound it so that it was now a single string of twenty-seven adjoining cubes. As you've probably guessed, the new team was about to be given a task that involved spatial rotation—the very skill about which Claire had just been praised. "Your job," Lisa said, "is to put this puzzle back together in the next five minutes." Lisa returned to the control room to videotape what happened next.

Ashley grabbed the puzzle first (as we'd instructed her to do) and worked on it for a minute. "Can I try?" Claire asked rather insistently, reaching for the wooden figure. Claire worked intently, but soon Maya wanted to give it a go. Only after Maya asked for a second time did Claire hand it over. As Maya took her turn twisting and turning the pieces to fit them together, Claire pointed and gestured to guide her; it seemed she couldn't refrain from offering advice on what moves to try. When Maya hesitated, Claire quickly reached for the puzzle, offering to take it. And so it went, not just with Claire but with all the other participants who received acclaim for their work on the original task. Overwhelmingly, the people we'd made to feel proud of their spatial abilities took charge of the group's efforts, monopolizing the puzzle for 20–25 percent longer than their teammates.[13] In many ways, this finding that inducing pride made people want to work longer and harder is just further proof of our earlier results. But what effect did this behavior have on how the others viewed them? Did the other members of the group respect them for taking charge, or did they resent them for being pushy and overbearing? To find out, at the end of the experiment we had all participants rate two things: how dominant they felt the others in the group had been, and whether or not they liked them.

First, we found that the people who hadn't received any feedback viewed their proud partners as having had significantly more say in the group's efforts, and as a result saw them as the group's leader and as having higher social status. So far so good. People's perceptions were shaped by what the proud participants did and how they looked. The most important question, though, remained: would these self-appointed leaders be liked and respected, or regarded as arrogant and annoying? To our pleasant surprise, the answer was clearly the former. Not only were they viewed as leaders, they were valued and admired for it. On average, those who demonstrated pride during the task were rated as significantly more likable by their partners than were those who didn't. And, significantly, their partners didn't just like them because the proud people did more of the work. Ashley, our confederate, always worked on the puzzle for much less time than either of the other two in each group, but she wasn't liked any less for loafing. No, the only factor that influenced how likable a person came off in this experiment was pride; pride made a person seem like a leader, and this in turn made that person a more desirable member of the group.

Taken together, this work shows that all it takes is a little praise or recognition to shift the scales of character; it not only motivates

The naked emperor

In Hans Christian Andersen's "The Emperor's New Clothes," a fashion-conscious despot is tricked into paying for a wondrous suit made from fabric so special (or so he's told by his tailors) that it is invisible to anyone who is unfit to hold his or her station in life. The tailors, of course, are swindlers. But the emperor is loath to admit that he cannot see the (nonexistent) clothes; to do so would be to admit that he is unworthy. So the emperor parades naked before his subjects, who at first go along with the charade; they don't want to admit to being unable to see the suit either. But after one child shouts the truth, that the emperor is in fact wearing no clothes, the proud emperor is soon outed as an egotistical phony.

We've all witnessed some version of this parable. And as we've seen, when a proud person's façade is shattered in this manner, they can go from loved and revered to reviled and derided seemingly overnight. It's one thing to project an air of pride and confidence, but if our skills and abilities don't justify our strutting, suddenly that pride becomes viewed as hubris—which sends an entirely different signal to those around us.

We suspect that the same would have happened in our experiment had it become clear that the puzzle-solving advice our proud people were giving wasn't especially helpful. Their leadership would be questioned, and their input would be seen not as expert assistance but as know-it-all bossiness. They would, simply put, be stripped of their status and likability. This is essentially what happened with Bush and Obama. Both project an air of pride and confidence, but whether they are celebrated as able leaders or dismissed as arrogant posers depends on whether the people doing the judging believe their performance backs up their strutting. Many Republicans believe Bush is a man of principle who kept the country safe, so to them, his pride is warranted. But the many Democrats who disagree see him as an incompetent ideologue—the poster boy for hubris. It's a similar case for Obama. If you believe he knows what he's doing, you love to hear him lecture the doubters. But if you believe he is still wet behind the ears, you see this as arrogant and off-putting. The same goes for Tom Cruise. Though the pride he exhibited as a young man led to masterly acting, once his career plummeted and he started pontificating about his religion and the meaning of life, the confidence and pride he projected weren't viewed as favorably. He had crossed the line to hubris.

So where exactly is the line between pride and hubris, and what determines whether we cross it? If hubris is so unlikable, why would someone overestimate his or her abilities or project pride for skills or achievements that simply don't exist? The answer is again to be found in the battle between the systems of the ant and grasshopper. As we've seen, pride has two main functions: to motivate us to persevere, and to signal our worth to others. The end result—high social status—is definitely a desirable thing, both in the here and now *and* in the future. The problem is, in the short term we don't always want to put in the effort necessary to earn this position. So what do our minds do? The grasshopper tries to lull us into a sense of false pride. It tricks us into scanning the environment for signs we excel at something even when we don't, and to take credit and exhibit confidence whenever possible, even when it's unfounded. In other words, it pulls us across the line into hubris.

Everything we've said to now seems to suggest that when this happens, it can only lead to trouble. But if that's the case, why is hubris still around? Why does the grasshopper, who like the ant strives to protect our interests, act as it does? It must mean that hubris can have some redeeming value. Yet if this is right, you're probably wondering where the examples of good hubris are to be found. The answer is: all around you.

The man behind the curtain

It was the last week of March 2009, and the United States was still in the grip of a massive economic meltdown. Companies were hemorrhaging jobs, the welfare lines were long, and unemployment was at its highest level in decades. Now, generally speaking, being unemployed is not considered a source of pride. Nor does it confer particularly high status in our society. In fact, it's quite the opposite; often it can lead to shame, so much so that many people will cop to almost anything before admitting to being out of work.[14] So you'd think that in March 2009, as unemployment hovered around 10 percent, countless Americans would be walking around with their heads hanging low. But they weren't.

As Ben Carey at the *New York Times* reported, something surprising was going on among the newly jobless. Many were still leaving the house in the morning dressed in business attire, commuting into town, and meeting with colleagues or business associates (or, more accurately, ex-colleagues and ex-business associates) for lunch. They'd then hold court at the local Starbucks, where they could get free Wi-Fi on their laptops. In essence, they were posing—they were living their lives as if their previous status were still intact. If you passed one of them on the street (and you probably did), you wouldn't see a despondent, unshaven slouch in sweats; you'd see someone who appeared to be among the relative few who survived the crisis—someone who was clearly valuable enough to his or her company, and to society, to have held on to a job even in the tough times. What Carey wanted to know was, why was this happening?[15]

Carey called our lab to see if we could explain the psychology behind this behavior that, on its face, didn't appear to make much sense. To us, however, the psychological mechanisms behind this posturing couldn't have been clearer—it was an example of the grasshopper at its best. These "Starbucks executives" were perfect examples of how projecting false pride can help us protect our social status, at least in the short run. Here they were, dressed in their office finest, typing on their BlackBerry with one hand and adjusting their Bluetooth headset with the other (maybe even using the fake-call app on their iPhone to seem in high demand). But where the kind of hubris shown by Tom Cruise, the fabled emperor, or the water boy who acts like he's the quarterback leads to disdain, for these folks it was actually a good strategy for success.

As we saw earlier in the chapter, physical expressions of pride automatically signal status and value. When we see someone who looks self-assured, who holds her head high, we assume she is important; after all, why wouldn't we? There is no equivalent signal for hubris—if there were, it would have disappeared long ago, as it would only serve to solicit ridicule. So in a brief encounter with someone about whom you know little, there is no way to tell if the pride and confidence he or she is expressing is justified. When a poser is strutting around looking important, we buy it. There is no way to know otherwise. And that's the point. By presenting the illusion of status and power, these people are positioning themselves to appear most attractive to potential colleagues and employers. Yes, it's an untenable tactic in the long run (like the emperor, they will eventually be found out), but in the short run it may provide an all-important competitive advantage that helps them get back on their feet. The grasshopper's gamble may pay off.

Hubris, then, can function as a protective mechanism. It can help us preserve our social status and, to some extent, our self-worth. This is why we often overestimate our abilities, sometimes subconsciously, sometimes deliberately. It's why, for example, as work by Richard Gramzow has demonstrated, we tend to misremember how well we did on tests such as the SATs, but with the important caveat that our errors go in only one direction—toward higher scores.[16] Gramzow has also shown that, just like the "Starbucks executives," people strategically present themselves in the workplace and other competitive settings to seem more accomplished and confident. What's most fascinating, however, is that this posing actually works to their benefit on many levels. Not only does it signal social value, as we've described above, but it has psychological and physiological benefits too, such as helping people stay calm during potentially stressful interactions. In one study, Gramzow and his colleagues Greg Willard and Wendy Mendes had participants take part in an interview while a computer monitored their cardiac responses. Amazingly, the researchers found that those who exaggerated their abilities in the interview actually exhibited less physical stress and anxiety than those who didn't, and as a result, they had a more successful interaction with the interviewer.[17]

Is such hubris a vice? We think not, but again, it's going to be context that is the ultimate arbiter. In the short term, the hubris of posing or assuming you're important can be hugely beneficial. If the gamble pays off, it can help get you into a position in which you'll eventually be able to feel authentically proud. On the other hand, if people see through you, you'll forever be branded a pompous fool. Remember, the ant and the grasshopper both want you to be proud.

But which opportunities they take to get you there and what they do with them are what tip the scales of character from one side to the other. Hubris can be useful for a short time, but on average the better strategy is to work hard to actually build the skills and achieve the goals that will make you legitimately proud.

5 / Compassionate or Cruel?

Looking into the mirror of the human soul

The winter of 1914 was not a happy one for the British and German troops manning the trenches outside Ypres, Belgium. World War I was in full swing, and their only breaks from exchanging gunfire were when they would venture carefully into the no-man's-land between their camps to retrieve the dead. It was cold. It was dark. It was violent. And the British High Command was intent on keeping it that way. They had very deliberately conditioned their soldiers to regard the Germans as bloodlusting psychopaths—the "evil Huns"—to ensure they'd have no qualms about taking as many lives as possible. And so, as Christmas approached, the opposing sides fought ferociously. But on Christmas Eve, something strange began to happen. As members of Britain's Berkshire Regiment peered across the fields, they started seeing small lights appear near the position of the Nineteenth Corps of the German army. At first they must have been frightened. Then they realized: the lights were candles the Germans had placed on small conifer trees in celebration of the holiday. Next the air began to ring with the sounds of German Christmas carols, to which the Brits replied by singing a chorus of their own.

Soon some of the men ventured into the space separating the armies and began to fraternize. Gifts of food and other small items were exchanged. It was hardly believable, even to those who were there. "If I had seen it on a cinematograph film I should have sworn that it was faked," wrote Lieutenant Sir Edward Hulse. Corporal John Ferguson was equally incredulous: "What a sight; little groups of Germans and British extending along the length of our front. . . . Where they couldn't talk the language, they made themselves understood by signs, and everyone seemed to be getting on nicely. Here we were laughing and chatting to men whom only a few hours before we were trying to kill."[1]

Upon hearing of this bizarre Christmas Eve détente, the British High Command was irate. How were they going to win a war if their soldiers suddenly became peaceniks? Accordingly, General Sir Horace Smith-Dorrien issued orders to forbid all friendly communication with opposing troops. He then instituted a policy of rotating troops so no one battalion could become too familiar or too friendly with their counterparts.

But they needn't have worried. The day after Christmas, the shooting resumed in earnest. Somehow, the soldiers suddenly had no problem whatsoever firing at the men with whom they had exchanged gifts and broken bread just a few hours earlier. What was going on here? How did these soldiers go from being mortal enemies to peaceful neighbors and then back to indifferent killers in less than twenty-four hours? It's difficult to understand. Imagine the cognitive gymnastics required by such a drastic turn of events. One moment you are willing to viciously take the life of another human being; the next you're sharing a beer and exchanging presents. You might chalk this up to the discipline of trained soldiers. Maybe the years they'd spent developing a stoic indifference to acts of physical harm enabled them to turn their aggression on and off at will. In other words, they're the exception, abnormal. No civilians would be able to engage in such violent aggression toward someone with whom they had recently forged a human connection, could they? Indeed, we might be wary of those who are able to turn their compassion on and off seemingly at the flip of a switch. We might question their character.

But the more we look around, the less anomalous this kind of behavior seems. Take the Ivory Coast, for example. In 2002, civil war broke out as the government lost control of the north to a rebel uprising. The violence between north and south continued for the next five years, with one brief interruption. In the early months of 2006, fighting between the rebel-held north and the government-controlled south was put aside as both sides rallied behind the national soccer team, which had just qualified for the World Cup. Incredibly, protests became celebrations as the team was held up as a symbol of national unity—that is, until the tournament was over. As soon as the final whistle of the final game blew, the two sides were back to killing one another. It was as though nothing had changed.

Similarly, in 1969, during one of the bloodiest civil wars in Africa's history, Nigerians agreed to a three-day cease-fire while Brazilian soccer legend Pele visited the country to play against local teams. For a window of time, hostilities were held at bay as all involved seemed suddenly able to shift their focus from the things that divided them to those that united them. Again, you might quibble about whether people are really able to turn their cruel instincts (or their compassionate ones, for that matter) on and off. Did the tension between these individuals really subside when the players took the field? Or were they just temporarily squelching their instinct to stand up and slaughter their cross-border rivals so as not to miss a moment of the game?

Humans' capacity for compassion and kindness is often underestimated (understandable, given the violence in the world). But consider the outpouring of caring and support in the days, weeks, and months after the attacks of September 11, Hurricane Katrina, or the earthquake in Haiti. People across the globe rallied together to offer the victims relief and prayer. Governments and individuals from warring nations temporarily put aside their differences to join in the relief effort, waving banners declaring "We're all Americans now." Yet after the initial shock of the disasters subsided and the photos of burning buildings, mangled limbs, and people trapped in rubble vanished from the headlines, it didn't take long for these same individuals to forget about the New York firefighters, the newly homeless of New Orleans, or the orphans in Haiti and go back to business as usual, even when there were still thousands of victims in need. How could all these people have displayed such compassion one minute, then be so cruelly indifferent the next?

That ordinary people—not just hardened soldiers—can shift so easily from cruelty to compassion and back suggests that these traits are more fluid than they seem. In fact, anyone who has ever lashed out at someone they care about in the heat of an argument knows the phenomenon we are describing. In your anger, you say something deliberately cruel or hurtful, but then the moment you see the pain register on the person's face or the tears well in his or her eyes, it's as if a switch flips, and your anger and frustration melt immediately into compassion and guilt. When we consider these situations, a soldier's ability to perform horrific acts of violence one moment and to shake hands with his enemy the next seems less like a disturbing character flaw than a fundamental property of the human mind. As we're about to see, the line between the psychological states that drive these drastically different behaviors does not appear to be particularly thick; we are equally capable of both types of responses, not just in times of war but in our everyday lives.

So what is it that determines whether we will be indifferent or caring, peaceful or violent, cruel or compassionate? As with other aspects of character, the potential for each lurks in all of us. Which one emerges at any give moment is, of course, decided by the struggle between our dueling systems. As we've noted before, both short- and long-term focused systems are absolutely essential for adaptive social functioning. Planning for the long term won't get you too far if you can't see the threat that's standing in front of you, and living only for the moment often won't do much for you past that moment. But when it comes to cruelty and compassion, exactly what is it that tips the scales from one to the other? What decides which side wins?

Mirror, mirror on the wall, who's the worthiest of them all?

A growing body of evidence suggests that an important factor underlying whether we show someone compassion or cruelty is the person's perceived similarity to us. It should take little introspection to realize that we feel the pain of those with whom we seem to share some commonalities. Countless studies have demonstrated that we not only consistently show more compassion to those we deem "like us," but that the mind makes judgments of similarity quite rapidly and spontaneously.[2] You can see how this plays out in a setting such as a battlefield, where the opposing sides hail from different nations or tribes, speak different languages, wear different uniforms, and stand on opposite sides of clearly drawn (both physically and ideologically) lines. Such differences are likely why it was so easy for the British soldiers in our example to bludgeon and shoot their German counterparts in the first place. Those dastardly Huns, the British reasoned, were nothing like *them*. Yet, in the light of the Christmas candles, this "like us"/"not like us" distinction got a little more muddled. Suddenly these enemies seemed more similar; they were fellow Christians who celebrated the same traditions and sang the same songs. And once they got to talking to one another, their differences receded even further. They weren't just "dastardly Huns"; they were husbands and fathers, just like the British soldiers were.

These same psychological mechanisms were at work in the case of the warring factions of the Ivory Coast, for whom it took the unifying force of World Cup soccer to allow them to see their reflection in each other. Same goes for the people who came out in droves to help the victims of 9/11, Katrina, and the Haitian earthquake—the crises shifted their focus away from all their squabbles and differences and onto their shared identity as human beings. But once the worst was over and they slipped back into their "us/them" mentality, their compassion swiftly abated. It only takes a quick glance at the headlines to see that most conflicts—be they national, political, religious, or personal—often come down to this very simple and automatic "like us"/"not like us" split.

How can such a basic distinction flip the switch between our most noble impulses and our most vicious ones? One answer lies in our evolutionary wiring. In the middle of the twentieth century, scientists were having a few problems explaining certain acts of altruism and compassion. After all, wasn't evolution all about selfishness and one's own survival? Once natural selection was understood to operate at the level of the gene, meaning that evolution not only favored behaviors that ensured we'd survive but also those that ensured we'd pass our genes to future generations, altruism toward family members was easy enough to explain. But what about the preponderance of evidence suggesting that we would also act altruistically toward perfect strangers? It was puzzling until the evolutionary biologist Robert Trivers proposed a theory for what he called reciprocal altruism.[3] According to Trivers, the motivation to act altruistically toward people with whom one shares no genes can be adaptive as long as there is a high enough probability that at some point in the future, these others will act altruistically in return. That is, if I scratch your back today, you'll scratch mine tomorrow, and then we'll both do better in the long run than we otherwise would have alone. The idea has since been used to explain how, over thousands of years of evolution, humans have become equipped with the capacity to care about the plight of and look out for the well-being of those around us. These tendencies have passed the test of time because they serve the adaptive function of helping us build those lasting social relationships that are ultimately critical to our survival.

But there's a tiny wrinkle in this logic. How can we identify *who* we can expect to scratch our back in return for us scratching theirs? After all, it's likely that not everyone would reciprocate our aid to the same degree. Plus, our resources are limited; if we were wired to be equally compassionate and giving toward everyone, we would have nothing left for ourselves. It isn't optimal—or even possible—to help everyone in need. So some psychological mechanism for picking and choosing is necessary. Enter our dueling systems. The ant (looking out for our well-being over time by building relationships) and the grasshopper (providing immediate protection from threats to our self-interest) together determine whom we help or hurt.

This is where the "us/them" distinction comes in. Similarity seems to function as this selective mechanism, signaling to us whether a person is someone we can rely on to reciprocate our kindness. In other words, similarity acts as a cue telling us that helping another person will likely lead to our own benefit down the line. It helps us answer the question of whether it's in our interest to exert efforts to help. Of course, like all other psychological mechanisms underlying character, this judgment is very flexible, and sensitive to association and context. It would need to be in order to explain how seeming enemies can go from fighting to socializing and back again in such a short time.

To see how flexible similarity is, consider the following example, often cited by the psychologists Gregory Murphy and Douglas Medin. Do you think a gray cloud is more similar to a black cloud or a white cloud?[4] If you're like most people, you'll answer black. Now, let's change the context. Do you think gray hair is more similar to white hair or black hair? Here most people answer white. Why? It all depends on the framework. For clouds, black and gray both imply dark skies and rain. For hair, white and gray both imply advanced age. What this means for our purposes is that perceptions of whether two things are similar can be quite dynamic. Gray is more similar to black, except when it's more similar to white. In essence, the similarity of objects can change at a moment's notice without our even intending it.

Though you may think impressions of another's character are quite different from judging the color of a cloud, they're really not. They, too, are formed immediately and are sensitive to changing environs. For Germans at war, the Brits were either men from a different country, in which case they were to be bludgeoned, stabbed, or shot, or they were fellow Christians, in which case it was fine to share some holiday cheer with them. For the rebels of the Ivory Coast, the soccer-loving southerners could be seen either as compatriots or as bitter political enemies. And after 9/11, the gun-toting Bible thumpers and the latte-slurping Ivy League elitists could turn to one another and see only their shared characteristics instead of their differences. When it comes to matters of compassion or cruelty, the weights tipping the scale shift more quickly than you'd think.

Judging a book by its cover

We are constantly sizing up the people in our social environment. What type of person are we looking at? What kind of information about this person's character can we glean from what they say, how they look, their clothes, their gestures, their expressions? That we use all these cues when deciding what kind of person is in front of our eyes should seem obvious. What may be less obvious is that for strangers, at least, we often make up our mind with a single glance. Think first impressions don't matter? Think again, because a host of psychological research conducted over the past decade has discovered that they do matter, very much.

As in the case of love and lust, it seems that we are very willing to make assumptions about people based on physical cues. Interestingly, studies have shown that not only do we make character judgments based on literally split-second exposures to people's faces, but these first impressions can actually be quite accurate. For example, a recent study at Tufts University found that we are surprisingly accurate at guessing an individual's sexual orientation after seeing a picture of their face for just 50 ms, while another study by the same researchers has shown that people are able to correctly predict the level of a CEO's success (as quantified by company profits) from brief exposures to their faces.[5] Similarly, judgments of politicians' competence based on nothing more than brief exposures to their faces can quite accurately predict their electoral success.[6] And split-second judgments about whether or not a soldier looks "dominant" can quite accurately predict the military rank he or she attains.[7] Taken together, these findings and countless others like them demonstrate not just how quickly we can size up other people but also the wide range of traits and qualities we can infer from a quick look. But what does this mean for how we decide who is worthy of our compassion? We might be able to quickly guess who's gay or straight, hardworking or lazy, a follower or a leader, but that doesn't tell us much about who is worth helping and who to avoid. For that we need to be able to make spontaneous and rapid judgments of another's value to *us*.

Jeremy Bailenson at Stanford University believes that such judgments may hinge strongly on whether we "see ourselves" in the other person—whether we believe they're similar to us. To examine whether or not this is the case, he and his colleagues decided to conduct a clever real-world experiment to see whether rapid intuitive judgments about similarity would affect people's voting behavior—in this case their votes for the governor of Florida. When Charlie Crist and Jim Davis announced their candidacy in the 2006 Florida gubernatorial campaign, they probably thought their chances of winning had something to do with their political savvy. Florida had recently been in the national spotlight due to the Terri Schiavo case, in which Crist, as the state's attorney general, had played a very public role when he decided not to allow the federal government to intervene in the decision to take Schiavo off life support. Could he use this publicity and his hard-line reputation to his advantage? Or could Davis spin Crist's efforts negatively, or perhaps highlight his own accomplishments in Florida's house of representatives, to wrest the governorship from Crist, who was somewhat more well known? Turns out none of these things would make much of a difference. In fact, according to Bailenson's findings, even the huge amounts of money spent on advertising and the hundreds of hours spent strategizing, campaigning, and kissing babies may have been a whole big waste of time, money, and energy.[8]

In the weeks leading up to the 2006 election, the researchers selected a random sample of people from all over the country to participate in a computer-based study. First, they were asked to upload a recent photograph of themselves (you'll see why in a minute). Then, the week of the election, they were shown a picture of each candidate and asked to complete a questionnaire asking them to indicate how they felt about the candidate on a host of measures. Now, they weren't given any other information about the candidates besides their pictures, yet they were asked to make judgments about how dishonest, moral, and kind the candidates appeared, as well as how the candidates made them feel, how likely they would be to vote for them, and the like. But here's the twist. Unbeknownst to the participants, the experimenters had used photoimaging software to morph participants' own photographs with the candidates' faces, using a ratio of 60 percent candidate to 40 percent participant, which was just subtle enough that the participants wouldn't be able to consciously detect the manipulation. So each participant had actually seen a real photo for one candidate and a hybrid of the candidate's face and their own for the other. What was the point? Bailenson and his colleagues wanted to know if making the candidates look more like the participants would be enough to change their judgments and preferences.

It was. Results showed that across the board, people had a stronger preference for the candidate whose photo was blended with theirs. No matter who the candidate was or what he stood for, the people rated the candidate whose picture had been morphed with their own as being more honest, moral, kind, and so forth—and they indicated they'd be more likely to vote for him. Now, you might think that since these people weren't all living in Florida and weren't going to be influenced by the outcome of the election, they simply didn't care about who won. That is, if they cared more about the outcomes or had more information about the candidate's respective positions, then maybe they wouldn't have been so influenced by trivial things such as facial similarities. After all, people should vote on substance, not appearance, when it really matters, right? Bailenson and his team wondered what would happen if the stakes were higher, as in a presidential election. A national sample might not know or care anything about the political differences between Charlie Crist and Jim Davis, but surely they'd know something about, say, the differences between George Bush and John Kerry.

So the researchers ran a separate study, this time blending people's faces with George Bush's or John Kerry's. Sure enough, the manipulation again had a significant effect on people's preferences.

Those who were strongly partisan one way or the other didn't budge from their opinions of the candidates from the previously held election, but independents and undecideds (those whose votes, let's not forget, tend to swing presidential elections one way or the other) showed a significant preference for the candidate whose photo had been morphed with their own.[9]

The extent, then, to which we see individuals as similar to ourselves, even on a superficial physical level, can have a huge impact on our attitude toward them. But while attitudes are one thing, actions are quite another. Yes, similarity leads us to value someone more, and maybe even vote for him or her, but does it really translate to going all out to help that person? We thought it would. To our minds, perception of similarity might be the key to explaining how an individual can go from being a compassionate altruist one moment to a callous aggressor the next. And we suspected that the age-old battle between the ant and the grasshopper was behind it.

Distress is in the eyes of the beholder

After Hurricane Katrina struck New Orleans, much was made of the systemic failure of the institutions responsible for providing aid to the city's victims. Citizens across the country began not only to condemn the quality of the response to the catastrophe at the local and federal levels but also to question both the care with which various federal agencies (such as the Army Corps of Engineers, which was responsible for the upkeep of the levees) had prepared New Orleans for a possible storm. Some more radical voices suggested that perhaps certain demographic characteristics of the victims contributed to the lack of preparation, as well as the lack of effort and urgency behind the emergency response. In other words, some speculated that maybe if the victims of Katrina hadn't been predominantly African American, the president and the Federal Emergency Management Agency would have done more to help. The singer Kanye West may have been going a bit far when he infamously claimed during a charity event that "George Bush doesn't care about black people," but his bold claim does raise an interesting question. When disaster strikes, do the subtle differences we perceive between victims and ourselves influence how much compassion we feel for them and, correspondingly, how much we are willing to help them?

There are few character traits thought to be more admirable than compassion. This makes some sense, as caring about someone else's welfare—being motivated to help other people when they are in need—is a crucial component in fostering the long-term bonds we need to thrive. At the same time, feeling compassion for every victim, as we noted, can quickly lead to less than optimal results. Not only would you quickly exhaust your resources, but you also would be incapable of being aggressive when it might well be needed. What tips the scales, then, to determine how we decide to use our limited resource of compassion? It's tempting to assume we base these decisions on relative need; that is, the more dire a person's situation, the more likely we are to come to his or her aid. But as researchers, we didn't think so. As you've probably gathered by now, we had to believe that it's not the nature of the problem that befalls others that determines how much compassion we will feel, but whether we see ourselves in them and their pain.

How were we going to test this idea? We'd need a way not only to present a situation where someone was in need but also to manipulate the similarity of this person to those in the position to help. If we really wanted to show that compassion is not a stable trait but rather fluctuates according to the competition unfolding between the ant and the grasshopper, we had to find a way to demonstrate just how sensitive our bleeding heart is to context.

As we thought about how to accomplish this, we realized that we had part of the answer already. Remember the experiments on moral hypocrisy from Chapter 2, where we had people observe our confederate, Alex, commit a moral transgression so we could see if they would judge him more harshly for it than they would judge themselves? Well, we decided to do something very similar, except this time we didn't care what the people thought of Alex. No, this time we were interested in what they thought of the victim, and whether perceived similarity would be enough to make people more willing to help him.

How many black bears live in the forest?

Steve and Phil strolled into the lab. Steve was an undergraduate participating in the experiment for course credit, and as far as he knew, so was Phil. But—surprise—Phil was actually a research assistant playing the role of a participant. The experimenter entered the room and told Steve and Phil that the first thing they would have to do was fill out a questionnaire that would categorize them as one of two personality types: someone who habitually overestimates or someone who habitually underestimates things. But in fact the questions we asked them had little to do with their personality, and we weren't interested in their answers to the questions at all. Who would be? We purposely asked them to estimate trivial things like the length in miles of the Massachusetts Turnpike, the number of black bears that lived in Massachusetts forests, and the height of the John Hancock building in downtown Boston. The goal was simply to create a sense of similarity or difference—"us" and "them"—based on a completely new and meaningless criterion. After answering the questions, Steve and Phil waited as their respective computers "calculated" the results (in reality they were random). Half the unwitting participants were told that Phil was an over- or underestimator just like them, while the other half were told that Phil was the opposite type of person. Nothing more. No "You two have the same taste in music" or "You both have a passion for American history." In fact, no personal information whatsoever was given. This was on purpose. We wanted to ensure that our manipulation of similarity was as trivial as possible.

The idea, of course, was to see if perceptions of similarity about something that didn't have any preexisting significance attached to it would have an effect on how much compassion participants would feel for a victim and how much they would be willing to help that victim. More specifically, how would Steve react if he witnessed a transgression against Phil? Would his actions vary depending on this arbitrary measure of similarity?

To find out, we brought back our bad guy, Alex (he was just so good at it), and had our real participant, Steve, play the role of the "secret watcher" described in Chapter 2. In other words, Steve secretly observed Alex as he assigned poor Phil to a long and difficult task while keeping the easy and fun task for himself. As in the earlier experiment, Alex had the option to flip a virtual coin to decide fairly who got what, but he didn't even try. Now came the important part. How much compassion would Steve feel for Phil, and how willing would he be to help Phil with his distasteful (and unfairly assigned) task? While Phil began working, Steve responded to a series of questions tapping his sympathies for Phil's predicament. Then, just as Steve was about to leave the experiment, the following message popped up on his computer screen:

> You have now completed the experiment. Please go to the experimenter to receive your credit. As you know, one of the other participants in the experiment has a long and difficult task to complete. It's not important to the experimenters who complete this task, it is just a quantity of material that needs to get done. So, if you'd like to help out in any way, indicate as much to the experimenter on your way out.

At this point, Steve could either hightail it out of the lab and go about his day or find the experimenter and offer to help. There was no social pressure since no one asked Steve to help—not Phil and not us. If Steve wanted to help, he had to take the initiative; otherwise, he could be on his merry way.

Now, if any of our participants did seek out the experimenter and tell him they'd like to take some of Phil's burden, they would be escorted down a long hallway and around a corner and placed at a desk, where they would be presented with a stack of about thirty math GRE problems. The experimenter would then tell them to do just as many as they wanted—whatever they didn't finish, Phil would complete later. Furthermore, once they were done, they were just to leave everything on the desk and take off. In other words, it was made clear that they would never again interact with the experimenter or with the person they were helping. So, in the case of people like Steve, any help that they did lend would have to be motivated by a legitimate desire to relieve Phil's suffering, as opposed to an attempt to gain any social rewards from the experimenter or Phil for the actions. At this point, the experimenter left them to their good deed, but secretly timed how long they spent working on this difficult task.

What did we find? Were our participants more likely to help Phil with the problems when they believed Phil shared this meaningless label as a fellow over- or underestimator? Not only was our suspicion correct, but correct to a larger degree than we ever imagined. A mere 16 percent chose to come to Phil's aid when he was dissimilar to them, but 58 percent—more than three times as many—chose to help him in the exact same situation when he was perceived as more similar. What's more, not only did many more people choose to help when Phil's "estimating type" matched theirs, they also spent significantly longer periods of time doing so than did the 16 percent who believed Phil was different from them but agreed to help anyway.[10]

It may seem a bit disconcerting that we were able to manufacture compassion and altruism in the lab with no more than a silly little tale about whether people tend to make similar types of guessing errors. But if this is all it takes to tip the scale of character one way or the other, it certainly goes a long way toward explaining how acts of beneficence or cruel indifference can fluctuate, even in the same person, in the blink of an eye. Still, one might argue that this experiment wasn't a perfect replica of a real-life situation. After all, the manipulation we used, though subtle, was explicitly defined. We told people who was in their group and who wasn't. Given how many social norms there are surrounding group membership (such as "take care of your own"), people might have just been responding in the way they thought they were supposed to act ("This guy is like me, so I'm supposed to feel bad for him and help him more"). We thought this unlikely, but just to be sure, we decided to rule out this possibility in our next experiment. To do this, however, we needed a way to make people perceive similarity on their own, without us applying artificial labels, and then show that this perception affected how much compassion they would feel and how much altruism they would display.

Moving as one

In describing his experience in basic training after being drafted into the army in 1941, historian William McNeill writes of the long, grueling hours he and his fellow soldiers spent marching single file about the dusty plains of Texas.[11] Such drills might seem idiotic to the onlooker. After all, if a group of soldiers were to march in unison during battle, a machine gun would quite easily mow them down. So why were drill sergeants so keen on having their soldiers practice day after day, until they were moving perfectly in sync with one another? It wasn't until later in life that McNeill began to see a possible reason for such an exercise. In hindsight, despite the heat and the fatigue, his recollection of the marching drills was one of pure enjoyment and camaraderie. The act of moving in time with others, he recalled, led to a "strange sense of personal enlargement"—a feeling of connectedness with those around him. Lest you think perhaps McNeill had been out in the desert sun too long, research has begun to suggest that moving in synchrony *can* actually make people feel closer together. It acts as a kind of social glue, binding individuals into a larger whole.[12]

This was just the feeling we wanted to create in our experiment. We thought the reason being physically in sync with another person forges a bond is because it makes people feel more similar to each other. If this were the case, then we should be able to make people actually feel more similar to another person (rather than us just telling them they were similar) simply by having them mirror the person's movements. No questions about black bears and no explicit labels would be needed. The mind's attention to synchrony would be enough.

Given this theory, we conducted more or less the same experiment as before, but this time we simply told people that the first part of the experiment involved rhythm perception. The idea was that we would make some participants tap their hands in sync with Phil, the "victim," while others would tap to a different rhythm. We suspected, of course, that the synchronous tapping would create enough of a sense of similarity to make participants feel more compassion and offer more assistance to Phil.

And indeed it did. First of all, those who tapped in sync with Phil readily reported feeling significantly more similar in personality to him on a survey than did those who tapped to the different beat. For reasons that they couldn't possibly articulate, simply moving their hands in unison was making our participants feel more connected. Next came the big question. Would this be enough to change the level of compassion people felt toward Phil? It was. Forty-nine percent of those who tapped in sync with Phil volunteered to come to his aid, compared to only 18 percent of the asynchronous tappers.[13] Plus, the more similar they felt, the more compassion they experienced, the more willing they were to help, and the longer they spent helping him with the onerous task.

The implications couldn't have been clearer. Feeling similar to another person appears to trigger our humanity. It signals to us that these are the people who likely will be there for us in the future, tagging a person as someone we should care about—someone we need homeless man probably needs food more than we need our morning Starbucks fix? By convincing us that the person in need of our help is not like us, or not even a human at all. Dehumanizing someone, stripping them of their identity as being capable of thinking and feeling and reacting as we do, makes it particularly easy to ignore and transgress against them. It seems almost unconscionably cruel, but there's a growing body of research suggesting that when we perceive another person as "not like us," this is exactly what we do.

History is rife with examples of this. The writers of the U.S. Constitution defined slaves as three-fifths of a person. The Nazis described Jews as "vermin," and the Rwandan Hutus described the Tutsi as "cockroaches."[14] Almost every time one group has treated another horribly, they've found some way of dehumanizing their victims. And while these examples might seem so extreme to suggest that dehumanization is confined to the realm of madmen and sociopaths, that assumption would be incorrect. On smaller scales, any of us are capable of it. Consider the following story.

On a cold December night in 2005, patrons were waiting in line at the Starbucks near Nineteenth and Cambie Streets in Vancouver, Canada. As the scent of smoke began to overwhelm the powerful aroma of ground coffee beans, most people inside kept chatting, quite indifferently. Outside the coffee shop, it was much the same, with several customers calmly shivering over their lattes and chatting on their cell phones as black smoke began to billow above their heads. Just one customer, concerned about the blaze, peered around the corner and noticed an unconscious homeless man wrapped in a comforter that had somehow caught fire. As the flames crept higher and higher toward the man's face, another customer, who happened to be a nurse from nearby St. Paul's Hospital, tried to wake him, to no avail. The nurse then tried to recruit others to help her get the man to a hospital, but no one seemed to care enough to interrupt their conversation or their newspaper. One woman turned to the nurse and said with disdain, "Just leave him alone, he's a homeless person. Forget it." Another said, "Don't call the hospital. They don't want *him*."[15]

Certainly this is a particularly egregious display of cruelty, and it's tempting to assume that these callous folks are exceptions, rather than the rule. But is their cruel indifference to the well-being of the homeless really that abnormal? Consider what you feel when you pass a homeless person in the street. Do you always give cash? Or do you assume that giving won't matter, or that your money will probably go to feed a drug habit anyway, and so you walk right past? What if that homeless person was seriously hurt or looked unconscious? Do you call the police, or do you keep on walking, knowing that you'll likely be late to work if you wait for an officer to arrive? Most of us have done both—sometimes we helped, but other times we didn't. How could we be expected to never do the latter? If you walk down a city block wanting to help every needy person you see, you wouldn't make it very far and your bank account would quickly dwindle. By the same token, it simply isn't possible to contribute to every charity, join every cause, or even expend mental energy feeling compassion for everyone who needs it.

Given this fact, we need some sort of mechanism that turns these feelings of compassion off, lest they completely overcome our lives. Dehumanization seems to do this. In order for us to absolve ourselves for our callousness, our inner grasshopper, in looking out for our pleasure and resources in the moment, tricks us (albeit subconsciously) into seeing dissimilar others as objects instead of human beings. The callous woman in that Starbucks most likely didn't consider the homeless person to be much different in kind from the comforter in which he was wrapped. When we see others as objects instead of fellow humans capable of feeling and experiencing the world as we do, their welfare becomes inconsequential. And if you think this phenomenon is limited to some coldhearted scrooges, we've got some bad news for you. The tendency to dehumanize seems to be a fundamental part of our psychology, and a necessary one at that.

In one particularly compelling study, Lasana Harris and Susan Fiske at Princeton University used brain scanning technology to investigate how people would respond to different kinds of social groups.[16] Interestingly, there are distinct areas of the brain that are activated when evaluating humans, and others that are activated when evaluating objects. While in the fMRI machine, participants were shown pictures of people from a variety of different groups: the elderly, the disabled, Olympic athletes, the homeless, drug addicts, the rich, the middle class, and so on. The researchers were interested not only in how people described their emotional responses to these groups but also in whether the subjects' brains would process images of certain individuals differently.

What Harris and Fiske found was somewhat shocking. When people saw images of those who belonged to what sociologists consider extreme out-groups (such as drug addicts and homeless people—those who we think are most unlike us), the social categorization areas of their brains (the ones that are involved in making judgments about humans) were quiet, while the areas involved in processing objects lit up like fireflies. Their minds, in essence, responded to these people not as if they were people but rather as if they were things. Even more surprising, this wasn't just an intuitive response. People actually reported strong feelings of disgust upon viewing the images of these outgroups, and, when asked to pick objects that best represented how they felt about the drug addicts and the homeless, they chose images like vomit and overflowing toilets. Moreover, the same areas of the brain that responded to the pictures of vomit and overflowing toilets responded to the pictures of the homeless and drug addicts. Given this, it's no wonder that many people, even those who at times seem the most caring, don't always feel the pain of and help those who may need it the most. When we perceive others to be so dissimilar from us, the parts of our brain that are responsible for treating others with humanity can turn off, allowing us, for better or worse, to numb ourselves to their plight.

Again, the evolutionary calculus behind this is simple. The less similar another person is to you, the less likely he or she is to care about your well-being and thus the less likely to reciprocate your kindness. The less likely the person is to reciprocate, the more appealing the urgings of the grasshopper become.

Consider the following vignette:

> Two days ago I broke up with my (girlfriend) boyfriend. We've been going together since our junior year in high school and have been really close, and it's been great being at FSU together. I thought (s)he felt the same, but things have changed. Now, (s)he wants to date other people. (S)he says (s)he still cares a lot about me, but (s)he doesn't want to be tied down to just one person. I've been real down. It's all I think about. My friends all tell me that I'll meet other (girls) guys and they say that all I need is for something good to happen to cheer me up. I guess they're right, but so far that hasn't happened.

When people read this kind of story they tend to express some feelings of compassion or sympathy for the person. But in one interesting study, a group of researchers led by Roy Baumeister at Florida State University found that if you made people feel socially isolated before exposing them to the story, it would decrease their sensitivity to the plight of those around them. To demonstrate this, they created a clever (though somewhat harsh) experiment. They had participants complete a bogus personality questionnaire and then told some of them that, based on the results, they were the type of person who most likely would not be able to develop any meaningful relationships later in life and thus would end up alone. Ouch.

Turned out that the people led to believe that they would become socially isolated did indeed care less about the plight of the girl in the story. Not only that, it also made them less likely to engage in any prosocial behavior in general, and even made them less sensitive to emotional and physical pain.[17] In short, it numbed them. It seems that when the possibility of developing beneficial long-term relationships is removed, either because the person in need doesn't appear to be the type of person who is worth your efforts (i.e., is dissimilar to you) or because you have reason to believe that you are unlovable and so your efforts would be fruitless, the scales tip toward the grasshopper and your impulse to care about the suffering of others switches off. If you can't count on anyone besides yourself, you might as well live only for yourself, right?

Again, this may seem as if the ant governs what's "good" (compassion) and the grasshopper governs what's "bad" (indifference), but remember, nothing is that cut-and-dried, and both classes of responses are absolutely essential to a successful social life. There are instances when seeing others as different may be necessary, such as when it protects us from being taken advantage of by others who would bleed us dry, from feeling others' pain so intensely that we are sad all the time, or from being unable to aggress in an armed conflict when much is at stake. By the same token, unfettered compassion can lead us astray; indiscriminately feeling for and helping all those around us is a one-way ticket to being the biggest sucker on the block and is simply not a tenable strategy over time. What's important to remember, though, is that your conscious mind isn't always the one that is making the decision of who is worthy of your compassion or of your disdain. The ability to turn our noble feelings on and off is a fundamental property of the human mind. All it takes is a lightning-fast assessment of another person to determine whether we'll care deeply about him or her or be callously indifferent to the person's misfortune—an assessment that can be guided by all sorts of seemingly meaningless contextual variables, with our character hanging in the balance.

6 / Fairness and Trust

The surprising elasticity of the Golden Rule

Mohammad Sohail was closing up his convenience store in Shirley, New York, much as he does every night. But the night of May 21, 2009, is one he will never forget. As Sohail was tidying up behind the counter, a hooded man menacingly brandishing a baseball bat burst into his store, demanding cash. But as Sohail dropped his hands below the counter, he wasn't reaching for the cash box. To the thief's surprise, Sohail pulled out a 9 mm rifle and pointed it directly at the assailant's head. The man, apparently fearing for his life, collapsed to his knees, sobbing. He pleaded, "I'm sorry. I have no food. I have no money. My whole family is hungry." In that moment Sohail suddenly saw the would-be robber as someone who needed help—help he could provide. Sohail asked the man to promise that he would never rob again, and when the man did so, he gave him $40 from the cash register and a loaf of bread.

Sohail then went to the back of the store to get the man some milk, but on his return the man had disappeared. Feeling good about having done a noble deed, Sohail assumed that would be the end of it. He was wrong.[1]

In early December of that same year, Mohammad Sohail was opening his mail when he came across a letter without any return address. As his eyes skimmed the page, he quickly realized the sender could only be one person. The letter read:

> First of all I would like to say I am sorry at the time I had [no] money, no food on the table, no job, and nothing for my family. [It] was wrong but I had [no] choice. I needed to feed my family. When you had that gun to my head I was 100% that I was going to die. . . . Now I have a new child and good job, make good money staying out of trouble and taking care of my family. You gave me forty dollars. Thank you for sparing my life. Because of that you change my life.

But the letter wasn't the only thing in the envelope. The reformed thief had also enclosed $50—the $40 Sohail had given him plus an extra bonus. Now, no one had been holding a gun to the man's head forcing him to repay Sohail's "loan." In fact, not even Sohail himself expected the money to be repaid, and certainly not with interest. Sohail's kind act—his taking a chance to help without any certainty that it would be valued or repaid—had not only brought about a major change in the recipient but also resulted in a profit. "When you do good things for somebody, it comes back to you," Sohail proclaimed.[2] It seemed the robber had decided to follow the Golden Rule.

What makes people go from stealing to repaying (with interest), or from acting selfishly to treating others as they would hope to be treated in return? Where does the Golden Rule come from, and what makes us decide whether or not to follow it? The answer strikes at the heart of what it means to be human. No matter where we live or what culture we belong to, human beings depend on the exchange of time, resources, and social support with each other. Whether it's asking a friend to help us move our furniture, a loan officer to give us a chance to fund a new business venture, or a neighbor to watch our kids for a minute so we can run down the street to the pharmacy, we rely on the assistance of others far more often than we think. Yet, even though we don't often consciously consider it, each of these transactions is characterized by risk—the risk that a favor or resources we provide won't be repaid, or that an act of generosity won't be reciprocated. Sure, sometimes the risks are greater than others—it's more costly when someone fails to repay a $1,000 loan than the ride we gave them to work—but there are always risks nonetheless. In the short term, it logically makes the most sense to take the money and run, so to speak. If you don't pay back the $40 someone loaned you or give back your time, you're ahead. As the saying goes, the one who dies with the most toys wins! But in the long term, this strategy is not a good one. If you act this way too often or to too many people, you'll get a reputation as a cheat or thief and be ostracized, and then no one will give you any "toys"—a big problem in terms of survival in the long run.

It's the classic battle between the grasshopper and the ant, with short-term gains on one side vs. long-term stability on the other. So what tips the scale toward the ant and pushes us to be fair even in the face of the temptation to do otherwise? One answer to this question is simple and straightforward: whether we're consciously aware of it or not, acting fairly is a strategic decision. We know intuitively that even though it would be great to keep the money, gifts, or other favors people give us without repaying in kind, if we don't repay, those favors will very quickly come to an end and we'll be doomed to die a cold and lonely death. But if this were purely a rational decision, certain behaviors just don't make sense. As our colleague the economist Robert Frank frequently points out, people are often fair and reciprocate the generosity of others even when there are no clear long-term benefits. For instance, he cites statistics that most people tip waitstaff the same amount when traveling as they do at their regular neighborhood haunts.[3] From a purely self-interested perspective, why would you do this? You'll likely never see the nonlocal waiter or waitress again, so there's nothing really to lose by stiffing them, whereas the server at your local joint might respond by spitting in your food, or worse, the next time you come around. Yet not to tip, assuming the service was fine, just feels wrong.

By the same logic, why would our would-be robber ultimately repay Sohail? He'd already made away with the $40, and even if Sohail had been expecting to be repaid, there's no way he could have tracked the anonymous thief down. What's more, the robber clearly wasn't a wealthy man. He surely could have used the $50 he sent back to his benefactor. So what rational reason is there for repaying a gift from someone whom you will never see again? Well, there really is no rational reason, but we never said the mechanisms that shape our character were always rational.

It is true that at a reasoned level, we can strategically gauge which actions will maximize our self-interest by weighing their trade-offs in the long and short runs. The problem with such analyses, however, is that we generally have incomplete information. For example, you might think you won't be in this restaurant again, so maybe you don't need to tip—but then again, your boss did say there was a small chance she'd send you back to this town for one more client meeting, so you can't be entirely sure. Or you're pretty sure you can ask your neighbor Jack for help moving your couch, as you'll be away on vacation when he is moving out of his apartment and might ask you to return the favor—but what if he decides to cash in by asking for your help with something else? The point is that these calculations, to the extent that we consciously make them, hinge on assumptions that may or may not be correct. Sure, these what-if scenarios are unlikely, but remember that while the conscious mind is quite adept at imagining many different scenarios, it often does so in a way that can suit our interests in the current moment (as we saw in the discussion of moral hypocrisy). The intuitive mind, at least when it comes to issues of fairness and reciprocity, usually decides more quickly and bluntly. It operates on simpler rules that don't involve weighing the likelihood of what-ifs. It doesn't take the time to create a story or rationalization. Which is why we suspect that when it comes to fairness and reciprocity, the systems of the ant usually have a leg up.

Why would the ant tend to be the winner at the intuitive level? We think it's because the risks of being selfish are simply too high here, even when a rational analysis would tell us we can get away with it. After all, being a cheat or a "taker" is one of the worst reputations anyone can have. Nothing leads to social isolation more quickly. And social isolation not only will make you miserable but also, evolutionarily speaking, will tend to drastically reduce your chances of survival. Which is why, on the intuitive level, playing fair and following the Golden Rule always feels safest in the long run, even if you can't quite see why at the time.

Paying back and paying forward: Gratitude and the Golden Rule

"Damn it! It's all going to hell!" This was a sentiment commonly uttered by Monica Bartlett when she entered Dave's office. Monica was a member of our team during her graduate school days and was the person who headed up some of the lab's most creative work on gratitude. Monica believed that gratitude was the psychological mechanism central to our seemingly universal desire for fairness. Gratitude, she'd argue, was what made us do the right thing and reciprocate kindness even when it rationally seemed there would be no benefits for doing so (or consequences for failing to do so). In this view, she had good company. The sociologist Georg Simmel referred to gratitude as the "moral memory" of humankind. The economist Adam Smith believed it to be a God-given moral sentiment that was necessary to ensure human cooperation. And even the biologist Robert Trivers, who, as we noted earlier, developed one of the most famous biological models of altruism, theorized that gratitude underlies many instances of cooperation. Now, these theories make sense—we've all been in situations where we couldn't shake that feeling of gratitude no matter what—but there is a big problem: no one had ever been able to prove them. Why? Because studying gratitude in an experimental setting is really hard.

As you might imagine, this issue—how to study gratitude and its effects scientifically—was usually the source of Monica's consternation. If we were going to show that gratitude triggers a gut-level, intuitive desire that can push any of us to be fair, we first had to develop a situation that would not only evoke gratitude but also let us isolate its effects. Herein lay the challenge. Making people feel grateful usually requires giving them a gift or favor. How could we know, then, if a person's subsequent actions necessarily were a direct result of feeling grateful? They might just stem from good manners or from the knowledge that people are supposed to reciprocate favors or gifts. That is, how could we show that actually feeling grateful to someone who helps you (as opposed to just feeling obligated to repay the person for helping you) is the engine that makes us act fairly? Solving this problem was going to require some creativity, but first we had to figure out how to make people feel grateful within the context of the lab.

We had tried several methods of manufacturing gratitude, but most ended in the same refrain of "It's all going to hell!" First we tried giving people small gifts, but this tactic usually didn't work, since finding a relatively inexpensive gift that most college students would appreciate enough to feel grateful turned out to be an impossible task. (Giving them beer would have been unethical.) Plus, we found that most students viewed any gift we gave them with skepticism that got in the way of gratitude; some saw it as compensation for participating, while others assumed it was part of the experiment and tried to figure out our true motives. So we scrapped the giving of gifts (it was getting expensive, anyway).

Next we tried a different tack: giving nonmaterial rewards (such as letting someone do something enjoyable instead of something onerous). We set up a situation where two individuals would arrive at the lab and be told that there were two different tasks that needed to be done. Much as in the moral hypocrisy studies, one of these tasks was short and fun, while the other was long and difficult. The experimenter told the participants that she was going to flip a coin to see who would be assigned to which task. Before she did this, however, one of the individuals (who was a confederate working for us) would stop her. "Hey, you know what?" he'd say to the true participant, "I have a lot of free time this afternoon, and you look like you're busy, so I'll do the logic problems." We had hoped that this would make our participants feel grateful. It didn't. It seemed the little voice in their minds didn't say, "Thanks!" It said, "Sweet! This other guy's a sucker!"

Now, you might wonder why the participants didn't feel at least a little grateful. After all, the other guy in the room had just done them a favor by agreeing to do the task they were hoping to avoid. But that's the important point: they were hoping they wouldn't have to do the long, boring task. They didn't know for certain that they would have to do it; there was still a chance they could win the coin flip. The simple fact that the assignment hadn't yet been given to them—that the problem wasn't theirs yet—meant gratitude was in short supply. Since it wasn't clear on either a rational or intuitive level who was going to get stuck with the bad task and, consequently, whether anything was really to be gained by feeling indebted to the guy who volunteered to take it, our participants didn't feel grateful; they reported feeling lucky. Many even said they viewed the other person as weak, or even slightly strange, for making the offer he did.

Now we were getting frustrated. We were beginning to wonder whether college students just might be immune to gratitude, or whether gratitude was simply impossible to manufacture in a lab. Finally, after more cursing and head scratching, we settled on a way to tweak the existing procedure in a way that we were certain would prevent participants from attributing their good fortune to luck. We were going to make them own a problem and eliminate any hope or expectation of escape—that is, until a "benefactor" swooped in and decided to take pity on them and help. If anything would trigger their gratitude, this should, because as we've seen, people intuitively know that success in the long run depends on repaying the generosity of others.

Feeling fair

"This sucks! When is it going to be over?" Pam wondered as she sat in our lab completing the third long—and we mean *long*—block of grueling word problems. We'd assigned Pam the rather unenviable task of looking at strings of letters flashing on the computer screen in front of her and deciding whether they constituted actual English words.

Baddax—no
Sinan—no
Cabinet—yes

Fun, right? And it got better. As Pam finally got to the end of what seemed an unending task, a message appeared on her computer screen:

Trials complete—calculating score

"Yes!" she thought. "At last!"—until, a second later, the computer screen went black. As far as Pam could tell, the computer had died, and with it all remains of her work. As she sat there alternating between fury and dread at the thought of having to start all over again, the other participant, who had been completing the same experiment in the seat next to her, was hurriedly getting up to leave. It shouldn't surprise you by now to know that this other person—her name was Allison in this case—was a confederate working for us.

"What happened?" Allison asked, feigning surprise as she noticed Pam's blank screen. "Mine didn't do that. I'll get the experimenter for you." As Pam sat staring glumly at her screen, Allison reentered with Monica, who matter-of-factly informed Pam that she'd call a technician to come and reset the computer, but that Pam would have to start over from the beginning. As Monica left to make the call, Allison looked at her watch and noted she was already late for her campus job. "But," she said to Pam, "this really sucks for you. Let's see if we can figure out what happened and maybe fix it." With that, Allison began rummaging around the back of the computer, pulling on cords, and hitting keys, all the while asking if anything she was doing seemed to help. Finally Allison surreptitiously hit a key that we had rigged to bring the computer back from its state of apparent death. When it did, a grateful smile broke out on Pam's face.

"That's it! Thanks!" Pam shouted.

When Monica returned to the room, there on Pam's screen were her three scores. There was no need to begin the drudgery again from scratch. With that, Pam spent a few minutes more answering some final questions about her feeling state while Allison left. Then, with the experiment over—at least as far as she knew, anyway—off Pam went.

On her way out of the building, however, Pam just "happened" to run into Allison, who had a clipboard in hand. Allison explained she was working as a research assistant for a professor who was studying problem solving. She said to Pam, "Look, I really need to collect some data. Would you be willing to help? There are probably more problems in this packet than you can finish, but completing any amount would be helpful." Even though it didn't sound very interesting or enjoyable (and believe us, it wasn't), and Pam was already pretty tapped out from all the tasks we'd just subjected her to, she agreed. After all, if it hadn't been for Allison, she'd still be redoing those computer tasks. Allison took her to a desk at the end of a quiet hall and gave her the packet. She told Pam to just do as many problems as she could and leave the packet on the desk when she was done. She then headed out to recruit more people.

"Great," sighed Pam, "word problems, letter mazes, logic problems—ugh." Yet, even though she could have left immediately with no one seeming to be the wiser, Pam still spent about twenty minutes working on the problems. How do we know? We were secretly timing her through a hidden video feed.

Now, someone might argue that Pam might have been an unusually fair-minded person. Surely most people wouldn't subject themselves to that kind of mental torture just because they felt grateful, would they? Well, yes. In fact, our results showed that the other randomly selected participants like Pam (those whose computers "crashed" and received help from Allison) not only reported feeling very grateful but also more frequently agreed to Allison's requests for help—and, what's more, worked 50 percent longer on the tedious problems than did participants whose computers didn't crash and who had no reason to feel grateful.[4] Okay, you're probably thinking: "Well, this still isn't much of a surprise; these participants knew they owed Allison for helping them a few minutes ago. They're probably just helping her because they know it's the right thing to do, Golden Rule and all." Fair enough. But what if we told you we also ran conditions of this study where the person down the hall who asked participants to complete the problem solving measure wasn't Allison but a complete stranger? Everything else about the experiment was the same. Allison was still the person who sat next to them and, in the cases where their computer crashed, helped them out. But this time, the person at the end of the hall with the clipboard wasn't Allison but someone they had never met. What would Pam and her compatriots do then?

In this case, there isn't really a logical reason to help the stranger. Why would you feel obligated to help someone if you don't owe them, let alone know them or expect to see them again? But some people decided to help nonetheless. As we suspected, it was precisely the people who left our lab feeling grateful to Allison who much more frequently agreed to help the stranger. In fact, the more grateful people reported feeling before they left, the more time they spent working at that lonely desk at the stranger's request.

What was going on here? On the face of it, it seems to make no rational sense. But remember, our rational brains evolved fairly recently, evolutionarily speaking, and in the old days, when our brains didn't have the capacity to work through a reasoned cost-benefit analysis to decide whether to help another person, decisions were made relatively simply and automatically. If we receive a favor, we feel grateful. And if we feel grateful, we pay back. But like all simple systems, this one can be tripped up. So when we quickly replaced a true benefactor (i.e., Allison) with someone else asking for help, the system hiccupped. It knows that feeling grateful means you should be fair and follow the Golden Rule; it just doesn't stop to check to see whom you're repaying. After all, most times when we're feeling grateful, it's the person in front of us we're grateful to, so why expend the extra effort to check? This system works most of the time, but on occasion it can be co-opted.

To show how easily gratitude can be misdirected, we ran one final version of this experiment. This time we ran three conditions.

The first two were the same as before: the computer breaks and the participant receives help (i.e., gratitude condition) or the computer doesn't break (i.e., neutral condition). In all cases, the person who requests help at the end is a complete stranger. The third condition was identical to the one meant to induce gratitude save for a single difference. As Monica was signing participants out of the session, she asked them, "So, that other person—Allison, I think was her name—helped you out by fixing that computer, right?" They invariably said yes, and then off they went, never suspecting that they were about to be met by a stranger asking for their assistance.

This simple question disarmed their intuitive systems in two important ways. First, it reminded people that their feelings of gratitude were bound to Allison; they couldn't now easily misattribute this feeling to the stranger. Second, forcing them to stop and think about how and why they were feeling grateful, and to whom, brought their rational systems back into the decision-making process. And indeed, the results showed that this simple act of reminding people to whom they felt grateful strengthened the hand of the short-term systems. This time, the people who felt grateful to Allison were no more likely to help the stranger than were those who didn't. Yes, they still felt grateful at the moment they were asked for assistance, but the rational forces of the grasshopper now had time and motivation to correct the intuitive hiccup of the ant to repay just anyone who crossed their path.

From this experiment, one thing is clear: whether we are fair partners and pay back our debts stems more from automatic feelings than from reason. We can always justify why we don't have to pay back just yet, but we can't help feeling grateful. More important, we are wired in such a way that our gratitude can be misdirected, leading us to repay our debts to the wrong person. The danger of this, of course, is that if we're feeling grateful, we're liable to help anyone who requests it. But in some ways this isn't such a bad thing. In fact, it can be quite adaptive if it doesn't happen too often, as it encourages people to take the chance on a stranger with whom they might end up having a mutually beneficial relationship.[5] In short, it's kind of like paying it forward, driven by emotion.

Still, this fact also makes us vulnerable to the ploys of others. Think about it. When is the best time to ask someone for a favor or for money? When they're feeling grateful (even if it's to someone else). Ever wonder why sometimes those charities asking for donations stick a dollar in the envelope or give you a "gift" of stamps or stickers that you never asked for? As the results of our experiment suggest, these tactics work. So the next time you're feeling grateful and you're tempted to do someone a favor, take a minute to stop and think about whether or not the person asking you for the favor is someone who really deserves it.

That said, most of the time gratitude serves a bigger and more important function in life than just upholding a quid pro quo. Gratitude doesn't only help us reap favors, acquire resources, or build wealth. It builds something that may be even more valuable over the long haul: loyalty and trust.

That warm feeling isn't just the Polartec: Gratitude as social glue

On the night of December 11, 1995, Aaron Feuerstein was attending a surprise seventieth-birthday party in his honor. Surrounded by friends and family, it must have been a wonderful celebration. Feuerstein was a successful businessman. He was owner and CEO of

Malden Mills, one of the biggest textile producers in Massachusetts, recognized worldwide as the company that makes Polartec fleece. But the night didn't end as well as it began. As Feuerstein sipped champagne and accepted birthday wishes, his plant thirty miles to the northwest was beginning to smolder. By eight o'clock, it had become engulfed in a six-alarm fire that burned for hours, destroying three of its most central buildings. It took more than four hundred firefighters to put out the fifty-foot walls of flame, and by the time they did, more than 600,000 square feet of manufacturing space had burned to the ground.[6]

Everyone expected Feuerstein to take his $300 million in insurance money and either retire or rebuild the plant overseas, where costs would be cheaper, as any shrewd businessman surely would have done. So on December 14, three days after the fire, when Feuerstein stepped to the microphone to address his workers and the local media, the majority of the crowd was hardly expecting him to announce holiday bonuses. Yet the news was far better than they could have imagined. Feuerstein was going to use the insurance money not only to rebuild the plant but to rebuild it right where it was. He was also going to use some of that money (in combination with bank loans he'd take out) to pay his employees their full salaries and benefits for the next ninety days, while the rebuilding began. This was no small gesture; salary and benefits totaled roughly $1.5 million per week. The more than a thousand people in front of him broke into hugs and cheers the likes of which are rarely seen at the site of a massive disaster.

Just three weeks later, enough of the factory had been rebuilt that 10 percent of the workforce was back on the job. By spring 80 percent were back to work. And once a year had gone by, the new replacement mill had been opened and any employee who wanted his or her old job back had returned. The new mill was bigger and better than the one that had been destroyed, but that wasn't the only difference. Productivity was way up. Before the fire, the plant had produced 130,000 yards of fabric a week. A few weeks after the fire, it hit 230,000 yards. And it wasn't just because of the new equipment and technology either. It was because the grateful employees were working longer and harder than they ever had before. "People were willing to work 25 hours a day," Feuerstein recounted. One after another, the employees described their gratitude to their boss. Angel Aponte, who had worked at Malden Mills for three years, typified the response: "The way I see it, there isn't anything Mr. A [as Feuerstein was known] could ask us that we wouldn't do. I even heard one of they guys say they'd take a bullet for Mr. A."[7]

It wasn't just his employees who were paying him back with their time and loyalty. As the story of Feuerstein's actions spread, he suddenly seemed to have spun a reputation of gold. Feuerstein had quickly become one of the most beloved men in America. NBC's Tom Brokaw called him the "best boss in America" and "a saint for the 1990s." Upon hearing that Feuerstein ate a dozen oranges a day, a Florida orange growers association sent him crates of fresh fruit. The Bank of Boston, the local labor union, and the Lawrence, Massachusetts, Chamber of Commerce sent donations. People all over the country—people he'd never even met—seemed to feel grateful to him too. He received hundreds of supportive letters from across the United States, many of which included cash ranging from a few dollars to $500. The gifts just kept pouring in.

On the face of it, this seems quite strange, as these people had nothing to gain by sending such gifts. Nonetheless, at the intuitive level, the rationale makes good sense. It seems that just as there are benefits to being fair and trustworthy, so too are there benefits to forging relationships with those we feel we can trust. It's obvious that we admire individuals, such as Aaron Feuerstein, who seem honest and who honor their responsibilities. These are the people that we want as partners and friends. When push comes to shove, we need someone to count on, someone who won't sell us down the river to turn a profit. As we've said before, social relationships are a two-way street. These potential partners also need to know the same about us. They need to know that our short-term interests won't always win, that we're in it to share both the profit and the perils. There needs to be some sort of social glue that binds people together.

We believed gratitude functions as just this type of glue. When those warm feelings of gratitude well up inside us, we feel so bonded to others—at least for the moment—that we become focused on our collective welfare and willing, like those Malden Mills workers, to make sacrifices for the collective good.

To test this theory, you can guess where we headed. This time, though, there wouldn't be months of "It's all going to hell." We had already developed a method to induce gratitude in real time. All we needed now was a way to capture the dynamics of trust and tease out its impact on the desire for individual vs. communal gain. Luckily, the field of behavioral economics offered a number of experimental "games" designed to examine these very issues. One of these games, known as the Give Some game, fit our needs perfectly. Here's how it works.

Imagine you're sitting at a table and the experimenter puts down four tokens in front of you. Each of these tokens is worth $1. At the end of the game you'll be able to exchange them for their actual cash value. In a different room, another person is sitting at a table with four tokens in front of him too. You each have the option to exchange any, all, or none of your tokens with the other person. Why would you want to do this? Well, because the rules of the game are such that when you do so, the value of each exchanged token doubles. That is, each of your $1 tokens is worth $2 to your partner, and each of his $1 tokens is worth $2 to you. This exchange, however, happens simultaneously. You won't know his decision before you make yours.

This set of rules nicely pits selfish against communal interests. If you were being totally rational about maximizing your own profit, you wouldn't give any tokens to your partner. That way, you're guaranteed a minimum of your original $4, and anything he gives you just adds to it up to a possible $12. What's more, you're only playing this game once, so he's not going to have a chance to punish you for being selfish later. On the other hand, the best communal outcome is for each person to give all four tokens to the other. This way, both of you end up with $8—double what you started with. Of course, choosing to give is risky, as it works out well for you only if you can trust your partner to do the same.

In several ways, this game mirrors many everyday dilemmas we commonly face. The difficult question, though, is how we decide whom to trust. What determines whether we follow the Golden Rule and go for the communal good or screw over the other person in hopes of maximizing our own profit? What would make someone loan money to a friend or help a neighbor with yard work and trust that those favors would eventually be repaid? What made the workers of Malden Mills agree to work harder to earn the exact same amount of money they would have received without putting in the extra effort? We suspected not only that feeling grateful would have something to do with it but also that, as we saw in our earlier study on gratitude and the Golden Rule, the effect of this feeling would radiate outward. People not only would be more trusting of those to whom they felt grateful but also would be more trusting of *anyone*. The ant would be dominant, at least for a time.

To test this theory, we ran our participants through the exact same procedures as before. They came to the lab, they did the onerous task, and their computer either crashed or it didn't. If it did crash, we again evoked gratitude in them by having a confederate help them fix the computer, thereby getting them out of redoing all their work. But then the procedure changed. We told them they were now going to complete a different experiment for the Behavioral Economics Lab. We then took participants to a separate isolated room and explained the rules of the Give Some game. We told them that the other person who was going to play the game with them was either the person they had just met in the lab or a different person who was a complete stranger. Among the people playing with the person they had just met, half of them were feeling grateful to him (as he'd helped them fix their computer) and half had no particular feelings toward him whatsoever (as their computer hadn't crashed). Same went for the people playing with the stranger: half were grateful to the person in the lab, half weren't. We next flashed some cash to the participants just so they knew we were ready to make good on our promised payments, then left them alone to decide how many tokens they wanted to give.

What did they do? Well, in the condition where they were playing with the person they knew, those who weren't feeling any gratitude gave about two tokens on average, whereas those who were feeling grateful gave an average of three tokens—or 50 percent more. So far so good. This fit with our prediction, but, as with the fairness experiment, it might be equally likely that they were giving more simply because they knew they owed the person who helped them, not necessarily because they trusted the other person to be generous. But now let's look at the condition where they were playing a stranger. As it turned out, people acted the exact same way. Grateful participants were more trusting than ones who were not grateful—they again gave 50 percent more money. These people couldn't possibly feel they "owed" their partner, nor could they have any idea about his or her trustworthiness based on past behavior, since they'd never met. Nonetheless, trust they did. It seemed that, in this moment, the gratitude we induced triggered intuitive systems that made people more trusting of everyone, and thereby more willing to take a chance for communal gain. It made them, in other words, behave like the kind of people you'd want as your friend or partner in the long run.[8]

This notion that gratitude helps to foster successful long-term relationships and social bonds has been borne out in a number of real-world studies as well. Psychologists Sara Algoe, Jon Haidt, and Shelly Gable have shown that gratitude among new friends is a strong predictor of whether they intend to spend additional time together.[9] When these researchers followed new pledges at a sorority where "big sisters" were assigned to give gifts to new "young sisters," they showed that feelings of gratitude for the gifts (irrespective of the size of the gift itself) were directly associated with how close new members felt to their sponsors. Similarly, work by Nathaniel Lambert, Margaret Clark, and their colleagues revealed that the more gratitude partners in relationships expressed, the more responsibility each person felt for the other's welfare. When they instructed participants to reflect on events that made them feel grateful to their friends, and then express these feelings to them over a three-week period, they found that both individuals attached more strength to the relationship.[10] In other words, gratitude made the bonds feel stronger.

A note of caution is important here, however. It may seem as if we're putting the systems of the ant up on a pedestal when it comes to issues of fairness and trust, and to some extent that may be true. But that's because, as we've noted, very little is as damaging to a person's social standing and success as to be branded a cheat or a welsher, and our long-term systems know it. Yet, as we've also noted, these systems, especially on the intuitive level, are not perfect. Their regular success depends on getting an accurate read of the situation and thereby making sure that when you're feeling grateful, it's to the right person and for the right reasons. Remember, we demonstrated that gratitude can make you more likely to agree to help anyone—not just your benefactor. And while this can be a good thing in limited doses, it can also get you into a tight spot if someone who is seeking to take advantage of you knows just when to ask you for a favor.

Similarly, the amount of gratitude you're feeling needs to be appropriate if the systems are to work well. At times some of us can experience gratitude too intensely. When this happens, we may misjudge how much we need to repay and open ourselves to big losses. We may work too hard or give too much to someone who helped us only minimally in the past. The intensity of gratitude we feel should be commensurate with the costs and benefits of what we received. If the equation becomes too unbalanced, we can run the risk of becoming a doormat. And, really, the same goes for all the other psychological forces that shape our character. Take guilt, for example. When experienced appropriately, guilt can keep us in line, but when it's experienced for the wrong reasons, it can also keep us in chains. For these reasons, we shouldn't dismiss the grasshopper as an evil influence. These systems looking out for short-term self-interest have to act as a balancing force on our decisions. They are what can prevent our long-term systems from turning us into suckers.

Monkey see, monkey do: Monkey don't see, monkey do even more

Honor students: the word brings to mind studious teens toiling for hours in the library. Honest kids who are putting in hard work now as an investment for a bright future, right? Well, that might describe some honor students, but unfortunately not the majority. A survey of more than three thousand top academic performers conducted by the folks at Who's Who Among American High School Students as part of their twenty-ninth annual poll revealed that 80 percent of the nation's best students have cheated on assignments to get where they are. These disturbing figures closely match the results of a similar survey by the Management Education Center at Rutgers University. What's more, when the Rutgers researchers asked students why they cheated, they plainly stated they believed it was just what those who wanted to succeed did. In short, honesty wouldn't cut it.[11] If everyone else was cheating, then it was a necessary evil.

This sad state of affairs demonstrates the important truth that the decision about whether or not to act honestly or fairly isn't always an internal one. External factors, such as whether or not others are acting honestly or fairly, clearly play a role. What's more, these decisions also seem to be swayed by whom the cheating is affecting: it's easier to cheat on something, be it a test or your taxes, when it seems the only one who might be harmed is some faceless entity, like a school or "the system." Why? Because when you don't have a face to put to the victim, you intuitively know that you're less likely to confront social consequences in the future.

Take the honor students as an example. Cheating appears rampant, and from what these kids say, it seems that they cheat in school because they believe it is commonplace, practically expected. They know that cheating in general is wrong, but in the specific instance of cheating on tests or homework, they do it nonetheless. One way to explain why this happens is to look at the old proverb "Monkey see, monkey do." Broadly speaking, there are two ways that people decide what constitutes correct behavior: what they learn and what they see. The problem is the two ways don't always go hand in hand. And which we choose to let sway us at any given moment often hinges on what is at stake in the long run.

Seeing is believing

To see just how malleable honesty is and how easily it's shaped by the honesty (or dishonesty) of others, Francesca Gino from Harvard Business School and her colleagues concocted a clever experiment.[12] They brought groups of eight to ten participants into a lab under the pretense of studying math ability. Each participant was given a brown envelope containing $10 in dollar bills and coins, along with a white envelope containing a score sheet. Each participant also received a packet of worksheets that contained twenty confusing math problems. Participants would have five minutes to complete the worksheet—not nearly enough time for anyone to possibly complete all the problems—and were told they would be allowed to keep 50¢ for every correct answer. At the end of the five minutes, participants were to record their own scores on a slip of paper, place it in the white envelope, and take the correct amount of money from the brown envelope.

Now for the really interesting part. The experimenters also varied whether or not the participants' answers could be checked, *and* whether there were cheaters in their midst. In one condition, participants had to bring their solutions to the experimenter, who double-checked their scores and the amount of money they took. In a second condition, no one checked their score—it was all on the honor system (the students were even instructed to destroy their work in a shredder, thereby removing any possibility that cheating could later be uncovered). In the final two conditions, the researchers made it clear that another person in the group was cheating. After about sixty seconds of working on the problems, a confederate would stand up and loudly proclaim that he had finished everything, which was obviously an impossibility given the time allotted. After shredding his work, he would announce to the experimenter that he was taking all the money from the brown envelope, as he'd solved everything. At that point, the experimenter, looking unsure but with no evidence to the contrary, told him he was free to leave. The only difference between this third condition and the fourth was whether this "cheater" was identified as attending either the same university as everyone else in the room or a neighboring one.

The results? As you might suspect, just the simple fact of being able to cover their tracks with the shredder—of not having anyone be able to see that they cheated—resulted in more cheating. Although there would be no logical reason to assume that participants in the different groups would have different mathematical abilities, those in the honor system group somehow seemed to solve more problems—about seven more, or $2 more, on average. But what was most interesting was that when the cheater was present, the number of correct solutions somehow magically went up again by about another $1.

What we see from this experiment is that cheating can be contagious, or as Gino and her colleagues put it, that a bad apple can spoil the bunch. It makes sense that more people cheat when the long-term consequences of cheating (i.e., someone finding out and sullying one's reputation) are eliminated. But how do we account for the fact that simply seeing someone cheat increases a person's own propensity to cheat? It seems that seeing that cheating is "normal" gives the grasshopper more license to urge you to take the easy way out. If others like you are doing it, it must be okay. Monkey see, monkey do.

What you can't see can hurt you

Sometimes, though, it's not what you see that makes you cheat; it's what you don't. So now let's take a closer look at the second psychological factor that often underlies cheating: the anonymity, or "facelessness" of both oneself, and the victim. In the battle between short- and long-term interests, anonymity has always been a friend of the self-serving side. Both literally and metaphorically, the shadows can conceal acts that you wouldn't be proud of but may benefit you in the short run nonetheless. Most of us would agree that, as we saw with the example above, the ability to obscure bad behavior, like cheating, often increases it. But how deeply ingrained is the link between visibility and deceit in the mind? There is evidence to suggest that the answer is: very deeply. So much so that making people literally less visible, simply by lowering the lighting in the room, can actually increase the prevalence of cheating.

Chen-Bo Zhong and his colleagues demonstrated this fact using a variation on the experiment with the math problems and payment envelopes we just described.[13] This time, though, there weren't any shredders or obvious cheaters. Participants simply placed their score sheets in one box and their worksheets in another before taking the money. In this version of the experiment, however, the researchers had a way to link reported scores to actual worksheets, even though the participants didn't know it. The other main difference in Zhong's design (which may seem trivial but, as you're about to learn, wasn't) was that some participants completed the math problems in a well-lit room while others did so in a dimly lit one. It was still bright enough for everyone to see each other, just not as well as if all the lights had been turned on.

When it came time to see what impact, if any, the slight adjustment in the lighting had on the amount of cheating, the researchers first made sure the actual math abilities of the two groups were equivalent by checking their performance on the worksheets. They were; those in the darker room solved the same number of problems on average as those in the brighter room. Honesty, however, showed no such symmetry. Those who had worked in the darker room overstated the number of problems they had solved correctly and pocketed almost $2 more on average than their honest counterparts.

But wait—the effects of anonymity get even more interesting. In a second experiment, Zhong and his colleagues had participants complete a common economic task known as the Dictator game. The rules are simple. You're given $6 to divide between yourself and another person. The other person is in a separate room, and this single interaction will occur anonymously over a computer. Once you decide how much, if any, of your $6 to give to the other person, he has the option of accepting it or rejecting it. His choice will not affect how much money *you* keep; it only affects whether *he* gets any cash. For example, if you decide to keep $4 and offer him $2, you will get $4 no matter what. If he accepts the offer, he will get $2; if not, he gets zip.

The seemingly fair thing to do would be to keep $3 and give $3 to the other. Of course, rational self-interest on your part would tell you to keep it all. Now, for the other person, the rational choice is clear: accept whatever is offered. After all, even if he's only offered $1, that's $1 more than he had when he walked into the room. Yet countless studies show that recipients often reject offers that they believe are too low and thus unfair. It's hard to ignore that intuitive feeling of being screwed over.

In his experiment, Zhong had participants play the role of the initiator, meaning they got to decide how much money they would offer to the other. Now here's the important part. Some of these initiators were given sunglasses to wear; others were not. Why sunglasses? Well, they obscure part of your face and make it seem darker, giving you a certain sense of cover, or anonymity. Amazingly, this simple difference was all that was needed to give short-term, selfish systems a boost. Individuals who weren't wearing sunglasses gave on average about $3, or half their money, to the other—exactly what you'd perceive to be a fair offer. The sunglass-wearing folks, well, they were a bit shadier. They offered about $1.85 on average—clearly a less fair, more self-interested decision. So while it's true that we cheat when we see cheating, it is also true that we cheat even more when we think we can't be seen cheating.

As we've seen throughout this chapter, then, our adherence to the Golden Rule can be quite elastic; the scales of fairness, trust, and honesty are always shifting. Simple, subtle cues stemming from either another's actions or the features of the environment can tip the scale strongly from one side to the other. Remember, it's not that the people who worked in the dim light or wore sunglasses were inherently more dishonest than those who saw brighter environs. It's just that this slight change in context caused them to act more dishonestly in that moment. Nor is it the case that Mohammad Sohail's robber was an inherently honest person who wouldn't dream of trying to get something for nothing and always repaid his debts (after all, let's not forget that he did try to rob the store in the first place). It's just that his lingering sense of gratitude pushed him to do what he instinctively knew was fair. In cases like these, it's both the changing nature of our interactions and situations as well as the imperfections, or hiccups, in our mental systems that determine how we'll act. One day we might decide to pay back someone and treat her the way we'd hope she'd treat us; another day we just might disappear with the cash in our pocket . . . perhaps to buy ourselves a new pair of Ray-Bans.

7 / Playing It Safe vs. Taking a Gamble

Risk, reward, and ruin

Harry Watanabe opened a small gift shop in Omaha, Nebraska, in 1932. His inventory was unique, consisting mostly of trinkets from Japan. His business, which he eventually named Oriental Trading Company, soon expanded into seventeen shops throughout the Midwest. Harry had two children, Terrance and Pam, and in keeping with Japanese tradition, Harry dreamed that one day Terrance would take over for him as head of the family business. In 1977 his dream was realized: Terrance become president of the company. Moving the focus of production to party supplies and favors, Terrance oversaw the rise of a business that would eventually serve 18 million customers, churn out 25,000 products, employ 3,000 workers, and earn $300 million.

You know the sort of trinket we're talking about: spider rings, rubber bouncy balls, key chains, and those miniature pink muscle men that expand when placed in water. Americans have been rooting through cereal boxes in search of just such prizes for decades. For us, these small plastic delights have been the surprise found in the hollow center of the traditional Italian *uova di Pasqua* (chocolate Easter eggs) passed around at our own family Easter celebrations for as long as we can remember. Indeed, a browse through the Oriental Trading Company's online catalog will surely be a walk down memory lane for any child raised in the last half century.

Terrance was incredibly devoted to the success of the company, so much so that his friends and family couldn't help noting how he was never able to maintain a close romantic relationship. How proud Harry must have been of Terrance, making such sacrifices to devote his life to the family business. How happy Terrance must have been to be the source of such pride. And indeed, how much pleasure Terrance himself must have taken in his own professional and financial success. How surprising it is, then, that in 2000, after shepherding the family business so responsibly for over two decades, Terrance sold the company and proceeded to blow through most of his hard-earned proceeds at Vegas casinos. We're not talking just a few thousand dollars. Terrance Watanabe lost a mind-blowing $127 million in a single year. Doesn't it seem quite strange that someone so successful, who had earned a fortune by making intelligent, calculated decisions about costs and benefits, could so foolishly fall prey to the lure of the flashing casino lights?[1]

You may be tempted to lump Terrance, who clearly had developed a gambling problem, in with all other addicts. And depending upon your views of addiction, you might see him as weak—unable to overcome the temptation of winning on the next hand. You might see compulsive gambling as a character flaw, signaling a type of person who is unreliable, untrustworthy, and certainly not the sort with whom one would want to conduct business. But is the person who loses $127 million at the casino so different from the person who plays the stock market or dabbles in real estate? The difference between these types of individuals, we'll argue, isn't so much their character—after all, all three of these activities are high-stakes gambling—but rather in their sensitivities to risk and reward. As we intend to show in this chapter, our perceptions of risk and probability can change on a dime and are subject to the push and pull of dueling forces in the mind. If you acknowledge this fact, then suddenly the woman pouring her weekly paychecks into the slot machines in Atlantic City might not seem as deviant or flawed as you think.

Consider an entirely different kind of situation involving a risk that most people have personally experienced: flying on airplanes. Surely you've heard the statistic that you're more likely to get into a fatal accident in your car on the way to the airport than to be killed in a plane crash, but for the many people who fear flying, this fact doesn't always provide much comfort. Though we may know that the probability of a plane crash is low, our intuitions are harder to convince. That's because when emotions run high, our assessments of probability and risk are skewed by all kinds of cognitive biases. For example, studies show that after a high-profile plane crash hits the headlines, people estimate the likelihood of being killed in a crash as being much higher than they might have the day before. Again, rationally, this doesn't make sense. The likelihood of dying in a crash on March 26, 1977, was almost exactly the same as the likelihood of dying in a crash on March 28, 1977, but it probably didn't feel that way to the many people who watched the footage of the 583 bodies being pulled from the wreckage after two 747s collided at the Tenerife airport in the Canary Islands on March 27. Indeed, much research has shown that simply being able to recall

But if you haven't guessed yet, we and other psychologists of our ilk don't put much stock in such models. Sure, it would be nice if decisions were made by making use of all available information and rationally weighing all the costs and benefits. If this were the case, then our decisions about whether to play the next hand, get on that transatlantic flight, risk taking that shortcut through a bad neighborhood, or skip the birth control this time would generally turn out all right. But unfortunately, the ant and the grasshopper are not truth seekers—each recruits all the psychological ammunition it can to convince you to go all in with a pair of twos, pop four Xanax to get you through the trip home, make that condom seem too far away to reach for, or have you reach for the Purell gel every time you shake a hand.

Risk and distance: Smelling the cookies makes them harder to resist

Food and sex. These are two things that are pretty much universally enjoyed. But they are also two things that consistently cause us to make errors in judgments of risk. Just as with gambling, when it comes to food and sex, what seem to be failures of will, like eating that second piece of cake or cheating on one's significant other, often actually boil down to our inability to accurately weigh the short-term rewards of our actions (e.g., satisfying a sweet tooth or a carnal urge) against the long-term risks (e.g., weight gain or ruining a relationship). When you're considering whether or not to add extra cheese to that pizza or buy those reduced-fat Wheat Thins, chances are you don't often stand there calculating the long-term risks involved in eating too much salt or fat, right? Similarly, if a partner tells you, in the heat of the moment, to forget protection and get on with it already, are you going to stop and rationally evaluate his or her sexual history? No. You have an urge, and you act on it. In other words, when it comes to food and sex, short-term pleasure seems to win every time. But are these urges to engage in risky behavior rooted only in our brains, or are they also sensitive to cues in our external environment?

That's what Peter Ditto and his colleagues at the University of California wanted to find out.[3] More specifically, Ditto and his team were interested in the extent to which people's sensitivity to risk hinges on the proximity of reward. In one experiment, they told participants that they would be playing a game of chance. If they won, they would get some freshly baked chocolate chip cookies that were waiting in the next room. If they lost, they would have to spend an extra thirty minutes filling out boring questionnaires. The rules of the game were as follows: Participants would pick a card from one of four decks of ten cards. Each card would be either a win or a loss. But different decks would have different odds of winning, and participants would be told these odds before they drew, at which point they could choose whether or not to play. The experimenters wanted to see how many people would choose to play the game at the varying levels of risk.

Now, if participants were at all sensitive to objective information about risk, then the results should be obvious: more people would choose to play the game when the risk of losing was lower (or the odds of winning were higher). And this is indeed what happened. But wait—the experiment wasn't over yet. Now the researchers wanted to see what would happen when the rewards stayed the same but were brought a little closer to home. So they conducted the experiment a second time. Here, instead of simply telling the participants that they could win cookies, they set up a small oven in their lab and actually baked the cookies right in front of the subjects. Would sitting in that room, with the cookies turning golden in front of them and the smell of freshly baked deliciousness wafting through the air, change their decision making? Yes. As the experimenter slipped on an oven mitt and pulled the hot tray from the oven to let the morsels cool ever so slightly, somehow the participants' willingness to take risks miraculously skyrocketed. As suspected, the temptation to gamble, even when the odds of winning were low, was now too much to resist. Participants' inner grasshoppers wanted those damn cookies, and they wanted them bad: "To hell with the possibility of consequences later! I have a chance to win chocolate now!" With this voice echoing in their subconscious, just as many people chose to play the game when the deck was stacked against them as when the odds of winning were high. It seemed, the researchers concluded, that making the reward more vivid and immediate can overwhelm the ability to weigh risks rationally. In other words, when the reward looms close, the risk becomes harder to resist.

Our perceptions of risk seem to be similarly swayed when we make decisions about sex. When you show men pictures of women and ask them to gauge the odds of contracting a disease from the women, the more physically attractive the woman, the lower the estimate.[4] When you think about it, this is completely irrational; after all, if anything, a sexier woman should be expected to have had more partners and therefore more opportunities to contract a disease. But we don't take the time to assess this logically when such an alluring immediate reward—sex with a gorgeous woman—is on the line. In a similar demonstration of the power of immediate rewards, another study showed that men who were sexually aroused reported being more willing to engage in risky sexual behavior than men who weren't aroused.[5] It's not that any of these men were inherently bigger risk takers; it was that when the visual and sensory cues of sexual opportunity are there, the desire for immediate pleasure takes over, turning even the most responsible guy into a carefree Lothario.

So the more appealing and immediate the reward, the more we instinctively ignore or downplay the risks involved. This may not seem particularly shocking. We've probably all been in situations where we're more than willing to throw caution to the wind in pursuit of something or someone we really wanted. But as we're about to see, when simple shifts in our environments completely blind us to the long-term consequences of our actions, it can have some pretty surprising—and often disastrous—consequences.

Risky business: Feeling your way

It turns out that many of the most important decisions of our lives, as well as the ones that seem to have direct implications for our character, are rooted in our subconscious assessments of risk. Whether he's aware of it or not, a smoker's decision about whether to quit is directly related to his belief about the odds that smoking causes cancer. Similarly, a voter's support for a policy geared toward, say, ending workplace discrimination and harassment will hinge on her judgment of how frequently these types of offenses occur. At first this may appear fine; after all, these people are grown-ups and free to make their own decisions. But the problem is, as shown by the studies described above, people rarely make these decisions rationally, although they like to believe they do. Rather, they allow emotional cues to override logic, which, more often than not, results in flawed decisions or judgment. We teamed up with colleagues Richard Petty and Duane Wegener at Ohio State and Derek Rucker at Northwestern's Kellogg School of Business to look at how small shifts in people's emotional states affect their assessments of risk and reward. If we were correct in thinking that simple changes in mood could alter your view of what awaited you behind the next door, the implications on our lives could be profound. For example, what if a smoker were suddenly less willing to go to a cancer screening because her good feelings about a recent job promotion made her underestimate the risk that the cancer would metastasize? Or what if a star athlete's elation after winning a big game inured him to the risks of unprotected sex?

To see how this might work, let's conduct a simple thought experiment. Imagine the scene during Hurricane Katrina. Think about the thousands of people desperately scrambling through the storm to find shelter, leaving their homes and, in many cases, their friends and family behind as they fought for survival. Picture those children who clung to their pets, only to be torn away by rescue workers and forced to leave the dogs and cats to certain doom. Think of the overwhelming grief of far-flung friends and relatives as they slowly received word of the loved ones they lost. Feeling a little sad yet? Now, answer this question: of the four million people in the United States who will propose marriage to someone this year, how many will be refused by the person they love?

This is more or less the exercise we put our participants through in our study. We had them read a news story that was intended to elicit a particular emotional state, such as sadness or anger, and then asked them to predict the likelihood of various other events. We overwhelmingly found that feeling sad or angry, simply from reading about an event such as a natural disaster or an anti-American protest in Iraq, was all it took to color their judgments about the odds of completely unrelated events occurring. It wasn't that hearing about an event such as a plane crash made them think plane crashes were more likely; it was that their emotional state swayed their general perception of the world around them. When people felt sad, they believed tragedy to be more prevalent; for example, they estimated that there were higher numbers of children starving in Romanian orphanages and brides being left at the altar. By the same token, people who were feeling angry overestimated the frequency of infuriating events, such as being screwed over by a used-car salesperson or being stuck in traffic.[6]

It may seem disconcerting at first to learn that not only do we fail to use logic when weighing probabilities but feelings and moods that have absolutely nothing to do with the decision being made can bias our judgments. But don't fret. It turns out that this tendency to overestimate risks can actually have its advantages, evolutionarily speaking.

Consider the following example: You're walking through the savannah with some of your family in search of a little breakfast. You come across a type of animal you've never seen before. It has dark brown fur with a white stripe down its spine. As you approach, it lunges at your merry band, sinking its teeth into your eldest daughter's neck and killing her. Now let's say we asked you what the probability is that the next animal with dark brown fur and a white stripe down its spine you see would be dangerous. You'd probably say 100 percent, and that's the most rational guess you could make since the single dark-furred, white-striped animal you've encountered proved to be dangerous.

Now, let's say you accidentally happen upon another one of

these creatures. This time the animal sits there peacefully, even assuming a deferential posture as you pass. Again we ask you, what is the probability that the next animal with dark brown fur and a white stripe down its spine will be dangerous? You'd probably pause. Rationally, your answer should be 50 percent, since as of this moment, one of two has proved dangerous. But your gut says something different. It's true that it is no longer reasonable to expect that all individuals of this species are dangerous, but on an intuitive level you know it's better to be safe than sorry. In your heightened emotional state, the cost of taking a longer path to avoid the brown and white critter is far less than the risk of losing another life. And in this case, your intuitive mind is right. While avoiding all animals with dark fur and white stripes would be an irrational calculation rooted in emotion (namely, fear), it is also an adaptive one.

Of course, this isn't just true in the jungle. In modern life too, listening to intuition and being more sensitive to the possibility of harm will serve you better on average than evaluating each individual situation rationally and objectively, particularly in situations that require rapid decisions for which you have incomplete information. It's hard if not impossible to know the odds involved in any given risk. What is the probability that you will get attacked if you walk down your own street? If you asked Kitty Genovese this question early on the night of March 13, 1964, she probably would have said it wasn't that high. But she *was* attacked. And she was killed. What are the chances you will get sick if you share a cup or if you eat a serrano pepper? Again, probably not that high. But tell that to the college students who contracted swine flu or fell victim to the salmonella outbreak of 2008.

The point is that our past experiences play a large role in our assessment of risk—perhaps an even bigger role than our mood or proximity to reward. When we undergo a painful experience, the desire to prevent such a thing from ever happening again can be so strong that we'd rather ignore the probabilities and just play it safe. If that means you have to avoid serrano peppers for a year, so be it. Our intuitive systems don't give much credence to that old maxim about lightning never striking the same place twice.

At the same time, having missed out on a reward in the past can make us more willing to take a risk in the future. For example, if you fold your hand in a poker game and the next card that's turned is the one you were waiting for, it's hard to convince yourself you made the right decision. Now the money you could have won is staring you in the face, coaxing you to go for it the next time and put it all on the line.

Studies such as ours have shown that not only does feeling sad or angry lead people to overestimate the prevalence of tragic or infuriating events, but feeling happy makes people more likely to overestimate the likelihood of positive events. This too is adaptive. How? Because it might compel you to take a chance on something you otherwise wouldn't have. Take a promotion, for example. Let's say only 10 percent of the people in your company get promoted to the next level. Logic and reason would tell you these are terrible odds and that you shouldn't even bother trying. But what if on one particularly sunny and cheerful morning your gut tells you just to go ask for that promotion even if, logically speaking, it's a fool's errand? What often seems like a fool's errand isn't, and if you put in the effort, you may just be rewarded. Sometimes you have to be in it to win it. So, it can often be better to listen to our intuition and play the possibilities than the probabilities.

But if following our intuition often leads to better outcomes in the long run, how does this explain Terrance Watanabe's gambling losses? It seems as though he had the opposite problem. Terrance wasn't in a situation where he had to make split-second decisions. The massive losses at the casinos unfolded over time. The answer is that Terrance was underestimating the risks. Instead, like the people who were more likely to gamble when they could smell the warm cookies, he was overly focused on the immediate reward. Each time he bet, the possibility that the next spin of the roulette wheel or the next turn of the card would win him the jackpot was so seductive, it blocked out all rational concerns about his long-term financial well-being or his family's reaction to his blowing their nest egg on a few rolls of the dice. When we think about judgments of risk and reward in terms of the battle between the ant and the grasshopper, Terrance's behavior and phenomena like it begin to make a lot more sense. The desires to avoid immediate losses and to obtain immediate rewards—whether in the savannah or in the poker room—all stem from the psychological processes geared toward our short-term interests. The processes that govern long-term interests are the voices in the back of our head advising us to forget about what's in front of our eyes and focus on what will be there much later on. And as we know, these are the voices that so often can be ignored.

So we see that gambling, or taking risks, is less about our "character" and more about situation and circumstance: our past experiences, our moods and emotions, and the visibility of rewards in that moment. The variability of all these factors is exactly what makes us seem to be daredevils one minute and straight arrows the next. When it comes to risk, our decisions are under the control of the ant and grasshopper, with important implications for how we are judged by those around us. In fact, understanding the processes underlying risk taking provides a compelling explanation for why we consider some types of people valiant heroes and others meek cowards.

tournament, the Butler Bulldogs knocked out a series of higher-ranked opponents on their way to a national championship show-down with the heavily favored Duke Blue Devils. It was painted as a David vs. Goliath matchup, and the nation was captivated. Even those who had absolutely no interest in college basketball were tuning in to see the drama unfold. Butler lost by two points after their last second shot clanked off the rim, but no one cared all that much about the outcome; the nation loved the Bulldogs for the mere fact that they'd gotten there by beating the odds. Taking risks that seem insurmountable may be the key to being seen as a hero.

Let's see how this psychological bias for the unlikely plays out in another competitive context: Wall Street. There is perhaps no group of individuals toward whom more vitriol and scorn have been directed over the past several years than Wall Street traders (or the greedy, callous, irresponsible, money-hungry leeches, as they're usually referred to). But it turns out that the psychological processes that cause us to root for the underdog (this attraction to beating the odds) might be the exact same ones that are responsible for the risky investment strategies that most likely contributed to the 2008 economic collapse.

Wall Street traders feel the same way about the high-stakes game of buying and selling that most people feel about sex and warm cookies: they like it. They like it *a lot.* To see just how much, Brian Knutson, a neuroscientist at Stanford, put traders into fMRI machines. Not surprisingly, when the traders were making high-risk decisions, the pleasure centers of their brains lit up like Christmas trees. And the riskier the decisions became (i.e., the worse the odds), the more pleasure they brought the traders.[7] In a way, the same thing is true for sports fans—the less likely the dark horse is to win, the more excited we are just to watch them play.

And the less likely it is that a firefighter will come out alive from a burning building, the more praise we heap upon him if he or she survives.

So the next time you curse those bankers on Wall Street and wonder at how they could possibly be so indifferent to the risks they were taking and the choices they were making, remember the pleasure you take in seeing Cinderella stories unfold. Sure, rooting for Seabiscuit doesn't have the same consequences as gambling away millions of dollars of taxpayers' money, but the psychology behind it is much the same. And, by the same token, the next time you're tempted to judge someone such as Terrance Watanabe for gambling away his family's fortune, remember that those same mental mechanisms that bring you so much joy in the fortune of unlikely winners are much the same as those that repeatedly drove him to bet thousands of dollars on a measly pair of twos. Again, when we look at risk in terms of the battle between the ant and the grasshopper, what seem at first glance to be deficiencies in character suddenly become a little more understandable after all.

Supermen and scaredy-cats

So if risk takers are heroes, then what about those who avoid risk at all costs? What about the cowards? To understand how common an aversion to risk can be, and why it is rooted in a fundamental property of the mind, let's first consider the little-known eccentricities of a famous figure: Charles Darwin. Darwin was nothing if not meticulous. And one particularly interesting detail about his travels that's not mentioned very often is the fact that not only did he keep detailed logs of the many species he encountered, he also kept a detailed log of his flatulence and bowel movements (as well as daily records of the severity and frequency of his tinnitus, or the ringing in his ears).[8] His writing on this matter certainly does not rank up there with *On the Origin of Species*, but it was something he evidently expended a considerable amount of time on. After all, he was known to be a hypochondriac. Hypochondria is a classic example of the human tendency to overestimate the possibility of immediate risks in our environment. If we were to ask Darwin or any other hypochondriac the likelihood that his stomach grumblings were symptoms of a serious ailment, he would most likely say close to 100 percent. Clearly, this would be inaccurate, but when our minds are always attuned to danger, we see it wherever we look.

This kind of mentality takes many forms. Agoraphobics confine themselves to their home because they've overestimated the risks they perceive in the outside world. Hoarders can't bear to throw anything away because they can't risk not having that old flowerpot when they need it. Of course, these are extreme situations, but in milder forms, risk aversion is actually an extremely common psychological trait.

Consider the following example. If we were to ask whether you'd rather have $50 right now or flip a coin for the chance to win $100, which would you choose? If you're like most people, you'd go with the former, and this makes sense. Though the expected outcome of each decision is the same ($50), there is risk involved in the coin toss—you might end up with nothing. But what if we asked if you'd rather have $40 or flip the coin for a chance at $100? Logically, if you calculated the risk, the odds of the coin toss would be in your favor, but you'd probably still choose the guaranteed $40.[9] This is an example of irrational risk aversion, also known as loss aversion, and most people experience it in one way or another. We seem to be wired to avoid immediate losses, even when it means sacrificing potential long-term gain. Yet as we've noted, in our culture this kind of behavior is often construed as a weakness in character. In fact, we reserve a word for those who avoid any kind of risky behavior: *cowards*.

Many of us feel like cowards at some point in our life. When we can't muster up the will to go talk to that person we've been eyeing all night, for fear of being rejected. When we'd prefer to keep all our money in savings accounts (or under our mattresses) so we don't lose it all in the stock market. When we refuse to walk home alone in the dark for fear of being mugged. When we don't let our kid eat that candy bar with the slightly torn wrapper in case it has a razor blade inside. These fears may not be rational, and they certainly aren't sexy, but again, they can be adaptive. In the long run, cowards are less likely to get rejected, lose their nest eggs, get mugged, and feed their kids razor blades. Which brings us back to the question at the heart of the chapter. What makes a person a risk taker in one context, and a coward in another? Once again, we see it has to do with our subjective understanding of the risks involved. Consider the child of a lifelong firefighter. Every day he sees Dad leave the house in the morning to go extinguish burning buildings and then come home safe and sound. Might this child grow up with a different idea about the risks associated with running into burning buildings than a child of a firefighter who died in a blaze? Of course he would. As we saw when we talked about the irrational fear of air travel, experience and exposure powerfully sway our perceptions of risk. So would the former child be more willing, later in life, to climb a fire escape to pull a baby out of a fourth-floor window than the latter child? Probably. But would that mean he's a braver person, a person of better character? A hero instead of a coward? Well, not necessarily.

The point is that "heroes" aren't necessarily braver people; they may simply have different estimates of the probabilities involved with the events. If you don't buy this, then you may have to reevaluate your opinions about adolescents, especially boys. Most people (over the age of eighteen, at least) would not agree that teenagers are necessarily more courageous or heroic than adults. But research has found that they certainly are less risk-averse.[10] Suggest to a fifteen-year-old boy that the two of you grab your skateboards and careen down the steps of city hall and he'd probably give you a high five, whereas most adults would look at you like you'd lost your mind. This isn't just because most adults look ridiculous on a skateboard. It's because the teen and the adult are wired to think differently about the risks involved. Research has shown that the teen brain hasn't fully developed the ability to develop what psychologists call "counterfactuals." In other words, they lack the cognitive ability to imagine the potential consequences of their actions (i.e., the skateboard going into the street and its rider getting flattened by an oncoming bus). And if a teenager can't even envision breaking his neck by skateboarding down a steep staircase, then how can he accurately assess the risk that it might happen? How could he be considered a hero for taking a risk he can't even fathom? So whether we act like heroes or cowards is not as much a matter of character as people tend to think it is. When the grasshopper is in charge, it can turn us into heroes, addicts, or cowards, depending on the context.

Tomorrow is always a day away

Imagine that on the table in front of you are four decks of cards. You know only two things about these decks. First, every card will have a number on it that represents the amount of money you will either win or lose, depending on the card. Second, the cards differ among the decks. But what you don't know is that in this game, known as the Iowa Gambling Task, some decks have better odds than others. The risky decks offer greater potential payoffs but have more "loss" cards; the safe decks offer smaller payoffs but at a more constant rate. But again, you know none of this, at least not yet. So how do you decide which deck to choose from?

When people play this game, at first they use trial and error; they pick from the different decks more or less randomly and see what happens. After about forty or fifty trials, however, they have developed a pretty good sense of which decks are safe and which are not, and then begin to choose cards almost entirely from safe ones. Why? They know that the game is going to go on for a while and therefore that their ultimate profit will be determined over the course of the game, not just on the next draw. In other words, somehow the systems of the ant kick in and shift people's attention away from short-term wins and onto the accrual of money over the long haul.

Intuitively this makes sense. Imagine playing the game again, but this time the experimenters have placed sensors on your skin that can gauge your arousal level by measuring increases in perspiration. That is, they can literally see you sweat. When Antoine Bechara and his colleagues did this, they found something fascinating: around the tenth card draw—long before you have any conscious inkling of which decks are risky—you begin to show anxiety (as measured by arousal level) each time your hand reaches to draw from what you will only later consciously realize is a risky deck. You're nervous, but you aren't even aware of it.[11]

This is a compelling demonstration of the ant at work. It acts as a silent statistician, calculating the risks and rewards associated with each deck and trying to steer you one way or the other based not on each individual flip but on the effect multiple flips will have over the long term. Left to the devices of the grasshopper, people might continue to flip from a deck from which they get immediate positive feedback or avoid a deck from which they've just been burned. But remember, the ant is focused on the *probability*, not the *possibility*, of rewards. After all, playing the probabilities is the key to success over the long term. The power of the Bechara study lies in its demonstration of just how subtly, how deeply below our level of consciousness, the ant can work. Clearly, we know on an intuitive level which of the decks are risky, otherwise we wouldn't be experiencing that anxiety. It takes us thirty more rounds—300 percent longer—to be able to consciously report this knowledge and adjust our behavior to minimize losses. Why? The grasshopper doesn't go down without a fight. The impulse to avoid immediate harms and gravitate toward immediate gains competes with the anxiety generated by the ant. In this kind of controlled situation, over time the scales tip toward long-term concerns and the players wise up. But in the real world, unfortunately, this isn't always the case. For many of the most important decisions in our lives, sometimes the ant needs a little help.

To your health

Earlier in the chapter we talked about how our perceptions of risk can impact health-related decisions such as whether or not to quit smoking or go for cancer screenings. In both cases we make these choices by subconsciously weighing the short-term benefits against the long-term risks. In the case of smoking, it's the pleasure of cigarettes vs. the risk of cancer. With the screenings, it's the reward of avoiding all that unpleasant poking and prodding (and the worry about receiving bad news) vs. the risk that a disease will go undetected. In their best-selling book *Nudge*, Richard Thaler and Cass Sunstein talk about how, by understanding the ways in which people think irrationally, we can help nudge them toward healthier, more responsible, and more productive behaviors.[12] Building on that idea, how can we use what we know about the psychology of risk taking to encourage people to be more responsible in looking after their health? In other words, how do we get people not only to hear the voice of the ant telling them to focus on the long term but actually to heed it? When it comes to our health, it's not enough to intuitively know those risks are there, like the players in the early rounds of the Iowa Gambling Task did. We have to actually act on them!

If you still believe that focusing disproportionately on risks makes you a coward, consider the following field experiment. Yale psychologist Peter Salovey was interested in how to get more women to go for mammograms. He quickly realized that in order to voluntarily subject themselves to the unpleasant procedure, women would have to judge the long-term risks of *not* going (cancer, possibly death) as being greater than the short-term costs of going (the hassle of going to the doctor, the physical discomfort of the X-ray, the mental anguish of worrying about a bad result, and so on). Logically, this seems like a no-brainer, but we shouldn't have to tell you at this point that logic has little to do with it. Manipulating mental and physical discomfort would be tricky, so Salovey and his team decided to focus on the risk part of the equation. They teamed up with a local phone company to recruit women in the New Haven area to come into his lab and watch short public service announcements on their lunch break. The announcements were of two types. Both urged women to get mammograms, but one video talked about the benefits of mammography (e.g., finding a tumor early increases survival odds); the other talked about the risks (e.g., not finding a tumor early can lead to death).

This seems like a trivial difference, but it actually turned out to have a huge impact on the women's decisions. Those who were made to focus on the long-term risks rather than the benefits were much more likely to later act responsibly and go for a screening. Why? Simple. When the announcement was framed in such a way that the ultimate long-term consequence was front and center, the ant suddenly couldn't be ignored. Here again we see how the gambles we take, even the big ones such as whether we're willing to risk our long-term health for short-term conveniences, can be greatly influenced by small and subtle differences.[13]

This may make it sound as if all would be well with the world if we always listened to the ant and focused on the long term. We might not have as much fun, but we'd be responsible and better off in the end, right? Well, that's true when it comes to our health, since the stakes are so high. But in other situations that rule of thumb doesn't always work because, as we've learned, the ant's foresight isn't always 20/20.

Ask any new professor what's the worst thing that can happen to his or her career and nine out of ten will give you this answer: being denied tenure. To avoid that future horror, they will make great sacrifices: working twenty-hour days, not spending as much time with their families as they'd like, letting their teaching responsibilities slide, and so on (trust us, we've seen it). But as work by Dan Gilbert and his colleagues has shown, all this extra effort may not, in the end, be justified.[14] Sure, being denied tenure is bad, but when Gilbert assessed the actual levels of unhappiness among professors who had been denied tenure, it quickly became clear that they were actually a lot happier than their younger selves would have predicted. And as Gilbert's team has shown, this type of prediction error for happiness is quite pervasive; we're as bad at predicting future happiness about all kinds of long-term rewards—everything from wealth to the outcome of an election and more—as we are at predicting risk. It's hard to make decisions regarding our long-term welfare if we can't accurately predict what will make us better off. Here again, neither intuition nor rationality always provides the answer.

So what does this all mean? Our decisions and behaviors are guided in large part by what our minds and circumstances trick us into believing about relative risks and rewards. Add to this the fact that our estimations of risks and rewards not only are very frequently flawed but are also quite fluid, and the mechanisms shaping character quickly become more complex. Once we come to grips with these dueling forces and how they can sway us—once we realize that we too are just one or two big poker wins away from a whole lot of more losses—then we can start making better decisions about when to gamble and when to play it safe.

3:40 Got a bunch of dead bodies lying there.
4:31 Oh yeah, look at those dead bastards.[1]

These words document the last minutes in the life of Namir Noor-Eldeen. As his name might suggest, Noor-Eldeen was Iraqi, but he was not an enemy combatant. To the contrary, he was one of the top Reuters freelance photographers documenting the American and Iraqi governments' efforts to root out insurgents in Baghdad and Mosul. This day, Noor-Eldeen, along with his Reuters driver, Saeed Chmagh, were taking photos in a Baghdad neighborhood where the Apache helicopter team was searching for insurgents. Noor-Eldeen had just snapped some pictures using his telephoto lens and was showing a few others the shots he had taken. By all accounts, he was calm even as the helicopter circled above him. After all, why should he worry? He was a photographer, not a militiaman.

The scene on the copter, however, was not so calm. The gunners had mistakenly identified Noor-Eldeen's lens and camera as an RPG and were circling to get him in their sights and take him out. By the time he realized that the copter gunners were aiming right at him, it was too late. Noor-Eldeen and his companions, none of whom had any weapons, were gunned down in a bloody massacre.

The release of this video (which the army fought for years to keep under wraps) has stoked much debate. If you listen carefully, you can hear the soldiers voicing hopes that the Iraqis will pick up a weapon (even though there weren't any there) so that, according to the rules of engagement, they could hit them with another round of machine gun fire or missiles. This series of events has led to public outcry against these military personnel. How could they mistake a camera for an RPG? With the level of training they'd had, how could they not recognize that Noor-Eldeen had done nothing to suggest he might be a combatant other than look Iraqi? The answer many have come up with is that these soldiers must simply be bigots—hungry for the blood of any and all Iraqis.

Although this view might seem tenable at first blush, on further analysis it really doesn't appear to hold water. The American soldiers often fought side by side with the Iraqi forces, sometimes putting their lives in one another's hands—not something you'd do with people you despised. Plus, this tragedy was just one isolated incident—a freak accident. If the soldiers were really prejudiced against all Iraqis, wouldn't there have been many more incidents like this one? And for whatever it's worth, the army's own internal investigation found no evidence of prior bias or an inclination to shoot before identifying the target. It was a tragic event but, at least according to the army, an unavoidable one that is part of the cost of war.

Still, when we read this story we couldn't help wondering whether the triggers would have been pulled so quickly if the man with the camera had been named Smith instead of Noor-Eldeen. If his skin had been lighter, if he had been blond, would there have been a little more hesitation, or at least a better attempt to verify whether what he was holding was in fact a weapon before the Americans opened fire? It's not that we believe the soldiers consciously shot the man just because they thought he looked Iraqi. But, as we've seen many times before in this book, what a person consciously thinks doesn't always dictate what he actually does.

Prejudice is one of the most reviled of human tendencies. Few of us would look at a bigot and say, "Now there's a guy with good character." Yet as psychologists, we can't help wondering: if prejudice is so bad, why has it stuck around for this long? As far as anyone can tell, stereotypes and prejudice appear to be as old as civilization itself. To have endured this test of time, there must be something that can sometimes be adaptive about them, something that, historically speaking, served a purpose, even if not a noble one. We realize this might not be the most popular argument in this book, but if you want to understand how to prevent bigotry from emerging, you have to understand the basis for why the mind engages in it in the first place. As part of this process, then, we intend to show that the question of whether prejudice is "good" or "bad" isn't always so (for lack of a better phrase) black and white. Which is why, as we will show you, when the circumstances are ripe, any of us, ourselves included, can act like a bigot no matter how fair and unbiased we believe our character to be. In fact, most of us, if placed in the situation of the soldiers on that Apache helicopter, probably would have acted similarly. Whether we like it or not, and whether we believe that prejudice is something we should all strive to overcome (which the two of us do personally believe), the human mind is wired for it—and this can influence people's behavior to an extent that you wouldn't believe.

I know I saw a gun

Imagine you're a New York City police officer scanning the neighborhood for a suspected felon. You see a man who might match the physical description and you begin to approach him. You are white. He isn't. As you're approaching, he turns to duck into the doorway of a nearby building. You identify yourself as a police officer, and as you do so, the individual reaches into his pocket and begins to turn toward you. You direct your gaze toward his hand, and you see he is holding a dark object. What do you do? Do you shoot or do you wait? You'd probably think that the answer most likely depends on whether the object he's taking out of his pocket looks like a gun. But that's not the whole story. You see, just how much that object (whatever it may be) resembles a gun depends a lot on who is holding it.

The notion that a mere error in perception can lead us to shoot an innocent man might seem (understandably) a bitter pill to swallow. Yet it's exactly what the psychologist Joshua Correll and his colleagues have convincingly shown in a series of inventive experiments that re-create the scenario above. Here's how it worked.

You sit down in front of a computer screen with two buttons in front of you. One labeled "shoot" and the other "don't shoot." The experimenter informs you that you'll see images of different street scenes flash on the screen in front of you—a city intersection, an alley, a parking lot, etc.—and every so often a man will appear in some of the scenes as well. The man will always be holding something—a wallet, a cell phone, or a gun. Your job is to "shoot" men who are holding guns by pressing the shoot button as fast as you can, just as you would if your life were actually on the line. If the man isn't holding a gun, you have to push the "don't shoot" button just as quickly.

We can all agree that if people took their time, no one would make any errors and no one would spot a gun where there wasn't one. A gun, after all, looks very different from a wallet. But we can also agree that when people need to identify the object in under a second, mistakes become a little more likely. But here's the kicker. Yes, Correll's participants made errors, just as you might expect given the time they had to make the decision, but their errors weren't random. Not by a long shot. His participants (all of whom were white) were much more likely to mistakenly identify a phone or a wallet as a gun, and therefore to shoot, when the man holding it was African American. The reverse pattern held when the man was white.[2] It seemed the participants' minds were engaging in some racial profiling on the intuitive level.

Now, these participants weren't bigots. They espoused no racial prejudices and had no history of acting discriminatory in any way. Yet here they were, deciding to shoot a black man much more readily than a white one. Sure, it was just an experiment, but the fact of the matter is that these same biases play out in real life. In fact, Correll based this experiment on a real-world tragedy you may recall from the headlines: the death of Amadou Diallo, a twenty-three-year-old Guinean immigrant to New York City. Diallo wasn't the criminal the police were looking for on the evening of February 4, 1999. He was an innocent guy selling wares on the street to make money for college. Yet as the police approached him, because they thought he might be the man they were after, he got scared and fled (as many in his situation might do), entering a nearby building. They ordered him to stop, and he began to turn around, reaching into his pocket for his wallet so that he could prove to them who he was. Unfortunately, however, the policemen were certain that the emerging wallet was a gun—a split-second mistake that resulted in Diallo falling to the ground with nineteen bullets lodged inside him.

But is it really fair to call this prejudice? In all these cases—the cops, the soldiers, the research participants—everyone *thought* they saw a gun. Wouldn't you shoot to protect yourself? Of course you would; almost anyone would if they thought they saw a gun. But that's exactly the point. Whether you *think* you see a gun isn't just determined by what's actually in front of your eyes. It's also influenced by the battle going on behind them.

each individual. It urges us to try to learn what he or she is like and not to jump to conclusions. The ant knows that to make a rapid decision about what someone is like based on the color of their skin or other marker can lead to missed opportunities. For the short-term systems, however, it's better to be wrong than to be dead. What matters most to the grasshopper is surviving right here and now, and given that the interests of different groups do often conflict, it may make sense to use a quick and dirty guess for what the person in front of us is likely to do. In other words, to use the only information we may have regarding a new person: stereotypes.

For better or worse (and often it's for worse), stereotypes provide the mind with a guess about what specific people are like. But if stereotypes are so bad, it raises an interesting question: why does the mind use them? The answer is simple: to help us make sense of the people around us. You see, stereotypes aren't inherently biased or maladaptive. They are just concepts that we use to categorize people in our social world, just as we use concepts to categorize objects in our physical world. For example, just as we know that chairs have four legs and are meant for sitting, we "know" that Italians are brilliant and irresistibly attractive. (What did you expect from two guys named DeSteno and Valdesolo?) In the absence of any other information, the mind uses these concepts to make predictions about new objects or people. For example, if we tell you something is a chair, you know you can sit on it even if it looks really strange (remember those chairs that were shaped like giant human hands?). Similarly, if you know you're going on a blind date with an Italian, you can expect it's going to be great. It's true that not all chairs have four legs and not all Italians are brilliant. But on average, if stereotypes are working correctly, most members of a category have the relevant features of the stereotype, and so our minds can use stereotypes as shortcuts to give predictive order to our world.

Now, while you may have accepted our chair example, you may have sensed some bias or self-interest creep in with the example about Italians. Our stereotype about Italians may not be the same as yours, which brings up an important question: how do we learn stereotypes in the first place? Usually it's in one of two ways: either someone plants an idea in our head about what people in group X are like (whether by telling us explicitly or by treating them certain ways), or we repeatedly see members of group X acting in specific ways and we extrapolate from that impression. For instance, back on the ancestral savannah, if every time you saw a member of the Mib tribe, they bludgeoned you, you would begin to avoid them at all costs, or to attack them before they hit you first. It certainly might be true that not every Mib would take a swing at you, but it might be safer to assume they would and avoid serious injury as opposed to taking a risk by conversing with them. Hence the potential benefit of stereotypes and prejudice. Of course, this strategy will not help you in terms of long-term peacemaking. Finding that one Mib who might well be interested in resolving hostilities between his group and yours could lead to great long-term benefits. But being wrong could also lead to broken teeth. Thus you see the contest between the two mental mechanisms playing out.

There is one more kink in the system, though. This is the one that often makes stereotypes so pernicious. Now that we're no longer on the savannah, what we see of group X can be very misleading. In these days of 24/7 media, what we learn about group X is often what the media decides to show us. If on any given day ninety-eight Italian men put in a solid day's work but one is indicted for being a mob boss and one commits a murder, which two stories will probably show up on the six o'clock news? The same goes for any other ethnic or social group. Unless we live in a cave, much of what we learn of other groups comes from the tragic or salacious stories we see on television. Back on the savannah, what we saw, we saw with our own eyes. So if Mibs were frequently violent, then the best guess the mind could make on encountering a Mib (at least in a statistical sense) was that he or she was going to be violent. But in today's sensationalist, media-saturated culture, what we see tends to reflect not the statistical realities but rather what is most "interesting" or aberrant. Yet the intuitive mind still uses that information to generalize.

This fact is why stereotyping people (as opposed to chairs) can be so problematic. Because our minds have been wired over thousands of years of evolution to take in small bits of information and generalize it to all members of a group, the usefulness of stereotypes can vary widely depending on the accuracy of the information. Even when we rationally know that all Iraqis aren't terrorists, or that all African Americans aren't criminals, or that all Italians aren't brilliant, our intuitive biases, irrespective of whether we endorse them, can shape what we think, what we see, and even what we do. If the officers confronting Diallo had waited an extra second or two—enough time to processes the information in front of them instead of just relying on intuition—they might never have made such a horrible mistake. What may be most surprising, however, is not only that our subconscious prejudices impact our behavior in unfair and dangerous ways but also that they have the potential to emerge seemingly from thin air.

Red, blue, I hate you

On the afternoon of April 5, 1968, the water fountain near Jane Elliott's third-grade classroom in Riceville, Kansas, was suddenly off-limits to students with blue eyes. Elliott had just told her eight-year-old students that blue-eyed people don't have as much melanin as brown-eyed ones, and that was important because melanin was responsible for intelligence and other good qualities. "Brown-eyed people are the better people in this room. They are cleaner and they are smarter," she said. "Blue-eyed people sit around and do nothing."[3] This was the pretext for one of the most famous and shocking examples of how quickly and arbitrarily prejudice can rear its ugly head. The evening before Elliott told her tale, Dr. Martin Luther King Jr. had been assassinated in Memphis. Now Elliott was desperate to teach her young pupils a lesson about prejudice. So she told them this fib about the superiority of the brown-eyed children. But it didn't stop there. She then proceeded to spend the remainder of the day praising the "brownies" over the "blueys," as she called them. It didn't take the children long to chime in. In just a few short hours, when a blue-eyed student got a math problem wrong on the board, the students said it was because he was a bluey. When a blue-eyed girl had to use a paper cup instead of drinking from the water fountain, a brown-eyed boy told his friend this was to make sure the brownies didn't catch anything.

This event is fascinating for several reasons. The foremost, though, is that it shows how readily the human mind—at least the young, relatively unformed human mind—will discriminate. These kids had all been friends. They had no history of any type of cliquishness or infighting. Yet all it took was an authority figure to give them a seemingly believable reason for why one group might be better than the other, and lines were quickly drawn in the sand. Suddenly even brown-eyed kids who were usually a bit quiet and timid were scoffing at their supposed inferiors. And to make matters worse, the exact same pattern of prejudice repeated itself on the next day—this time in the opposite direction—after Elliott informed the class that she had made a mistake: it was less melanin, and therefore blue eyes, that was associated with desirable qualities. Now it was the "brownies," according to Elliott, who were inferior. And the class bought right into it.

This demonstration was one of the first—and most resonant—to suggest that the capacity for prejudice lurks within everyone. Yet on the face of it there are several reasons to suspect that this view of character is too dismal. First, these were little kids, and kids are impressionable. They will believe whatever you tell them, especially if the person doing the telling is an authority figure. These kids, then, probably accepted the "facts" about melanin and eye color because their teacher told them it was true. Similarly, they discriminated against the blueys or brownies because their teacher did. So there's no reason to think that adults would ever act this way, right?

We decided to find out. In this case we were joined by Nilanjana Dasgupta from the University of Massachusetts at Amherst, one of the foremost experts on the fluidity of prejudice. If what we all suspected was correct, under the right circumstances prejudice could emerge in anyone. And if prejudice is really a function of the battle of mental systems, like so many other aspects of character we've discussed thus far, it should crop up even if you don't have any preexisting stereotypes or biases about the group in question. No explanations for why one group is more worthy are needed. No authority figure telling you how to behave is necessary. We suspected that readiness to discriminate is so ingrained that only two elements are required: knowing that someone is different from you, and being in a situation that amps up your inner grasshopper. In our case, we decided to manufacture such a situation by making our participants really, really angry.

"Okay, I tend to overestimate things," Michael thought. "Who cares?" Michael had just completed a seemingly innocuous questionnaire, similar to the one that we used to manufacture similarity in the experiments on compassion described in Chapter 5. It was simply titled "General Knowledge" and consisted of questions like "How many flights take off from Logan Airport on a given day?" and "How many miles long is the Massachusetts Turnpike?" Just to remind you, these were questions to which we assumed our participants wouldn't know the exact answers, and we were right. We simply told them to provide their best estimate. After each of them had done so, the computer in front of them churned for a while, appearing to calculate their scores.

In actuality, though, here again the computer was randomly deciding which of two responses it would give individually to Michael and his peers: "overestimator" or "underestimator." As you'll recall from Chapter 5, where we used these same labels to study the impact of similarity on compassion, these were simply meant to create two groups of people based on a ridiculously trivial and arbitrary difference that had no preexisting ideas attached to it.

"What's this?" Michael asked as Aida, our assistant running this study, handed him a red wristband.

"It's to mark you as an overestimator," she replied. "Everyone who is an overestimator will wear a red one. Underestimators will wear a blue one."

"Fine," Michael replied, very possibly suppressing an eye roll. "Let's get on with it."

Next Aida told the participants that two parts of the experiment remained. First we'd be assessing their memories for events, and then we'd be measuring their hand-eye coordination. For this first part, either they would have to describe an event from their past that made them feel very angry or they'd have to describe their daily routine.

"Something that made me angry, hmmm . . . ," Michael mused as the screen in front of him indicated he was to describe an angering event. After a moment's reflection he began typing, reliving the anger more and more with each keystroke.

"Time's up!" Aida said after a few minutes. "Now it's time for hand-eye coordination. Follow the instructions on your computers."

On the screens in front of them, each participant saw alternating images of words and people. The words could be easily categorized as "good" (*love*, *beauty*) or "bad" (*vomit*, *disease*), and the people would be easily categorized as "overestimators" (people wearing red wristbands) or "underestimators" (people wearing blue wristbands). The goal of the task was to make these categorizations as quickly as possible using just two keys. One key was to be used for both "good" and "overestimator" and the other for "bad" and "underestimator." However, to make the hand-eye coordination task (which of course it wasn't) even more challenging, halfway through the task the fingers used to categorize the people would change. So the key that originally meant "good" for words and "overestimator" for people would still be used to categorize "good" for words but all of a sudden would be used to categorize "underestimator" for people.

If this task sounds somewhat familiar, it's because it was another version of the Implicit Association Test (IAT) that we earlier noted is used to measure intuitive associations. This time, however, we were using it to gauge the mind's automatic response to other people, not to participants' evaluations of themselves. It's a task that has been used hundreds of times as a way to assess prejudices that may lurk below people's consciousness. As we said earlier, the mind automatically makes snap judgments of whatever we focus on. But it doesn't just categorize these people or objects; it also evaluates them as "good" or "bad." The IAT is designed to measure such snap evaluations by examining how quickly those judgments can be made. The basic idea is that you're quicker to categorize an object if you feel similarly about the one you categorized previously using the same finger. For instance, you'd be quicker to categorize a cute baby by hitting a key with your right finger if the word you saw before it was also good (e.g., *love*) and you categorized it also using your right finger. If you had to use your left finger for "good" words and your right finger for "bad" words, you'd be slower to categorize the baby with your right finger (assuming you like babies).

In the present case, this means that if the mind doesn't value overestimators any more than underestimators, it shouldn't be any more difficult to categorize one or the other depending on which key it's paired with. But if a prejudice exists—if the mind has a knee-jerk negative response to one group or the other—then the time to categorize them on the IAT task with the same finger used to categorize "good" words slows down. Let's go back to Michael and you'll see what we mean.

"Damn it," Michael muttered. He was having a hard time of it. When the keys switched, he kept making mistakes, and that slowed him down terribly. Every image of an underestimator that flashed before his eyes was accompanied by a feeling like a pit in his gut. He couldn't put his finger on it, but it slowed him down when he had to categorize them using the same key as he had just used to categorize the word *beauty* as "good." His gut, it seemed, was not fond of underestimators.

What does this say about Michael? Was he a bigot? How could he be? It would seem impossible to be prejudiced against a group of people that you really know nothing about, right? Why would he hate underestimators? Or why would others hate overestimators, for that matter? It's a silly distinction with absolutely no consequence. And for many of our participants it remained that way. Their response times on the IAT didn't differ as a function of which keys were paired. But here's the catch. Remember how the first part of the experiment asked people to either recount something that made them angry or describe their daily routine? Well, we did that because we wanted some of our participants—Michael among them—to feel angry, so we could see whether being angry would bring out any signs of prejudice or bigotry. It did. Those, like Michael, who were feeling angry had much longer response times when characterizing people who belonged to a different group when the response key was paired with the one also used to categorize good words. Those who described their daily routine, and hence were feeling nothing in particular, didn't show any such bias. As we suspected, simply being angry was all it took to create a prejudice from thin air.

To understand why this happens, let's think back to the ancestral brain for a minute. Back in the days of tribal competition, when violence was imminent, who was more likely to be the culprit, "us" or "them"? Hands down it has to be them, whoever the "them" are. So if a person sensed that aggression was likely (which would be signaled by the feeling of anger), the short-term systems of the mind went into red alert and "profiled" using the only criteria they had—better safe than dead. So the grasshopper, gaining precedence for the moment, makes you a momentary bigot. Fast-forward to our lab again. When the angry participants saw a guy from the other group—in other words, one of "them"—they instinctively hated him. Those who didn't feel angry felt no threat, so their intuitive systems had no reason to judge these others in a negative manner.[4]

The most important point here is that a simple change in context—feeling angry, for example—can cause prejudice to seemingly come out of nowhere. What's more, it can do so instantly and arbitrarily. Perhaps most unsettling of all, it can direct this prejudice toward people who, rationally considered, pose no threat whatsoever. Just ask Mel Gibson.

Mel, as many of you know, was widely hailed for his work as an actor and director during the 1990s and early 2000s. His philanthropy was well known, and for a time his "piety" was inspiring to many. Lately, though, Mel's character has seemed quite in flux, as he seems to go from generous and devout Catholic to racist pig and back again. Let's look at a few examples. Mel has been caught on record uttering homophobic remarks, yet he joined with the Gay and Lesbian Alliance Against Defamation (GLAAD) to host ten gay and lesbian filmmakers for seminars on the set of one of his movies. When stopped by a police officer in 2006 for speeding, an angry Mel, believing the officer was Jewish, muttered, "Fucking Jew . . . The Jews are responsible for all the wars in the world."[5] But right afterward, he voluntarily met with Jewish leaders to apologize for his "moment of insanity" and seek guidance on how to heal. And in the most shocking event to date, Mel was recorded by his ex-girlfriend Oksana Grigorieva screaming at her that she dressed like a whore and if she were "raped by a pack of niggers" it would be her fault.[6] Yet Whoopi Goldberg, a longtime friend of

Mel's, continues to attest that he doesn't have a racist bone in his body.[7] True, it's easy to dismiss Gibson's attempts to redeem himself as nothing more than PR stunts, and maybe they were. But that's not really the point. The point is that Mel was capable of such wild swings in behavior in the first place—that alongside his prejudices, some sort of social conscience must have lurked.

The reason these events seem so hard to put together is that, in the old view of character, they make little sense. Is Mel a bigot or isn't he? But when you think about it within the framework we suggest, the right question isn't whether Mel is a bigot but rather why he clearly acts like a bigot sometimes but not at others. The answer: context. Just as with our participants, though to an admittedly much greater and more repulsive degree, anger seems to have been the psychological mechanism that triggered the emergence of Mel's latent prejudices.[8]

Up to this point we've been defining prejudice simply as the mind's intuitive bias against one group or another. That may be scientifically interesting, but if we truly want to show that changes in these psychological forces impact character, then we need to convince you that these subconscious biases actually do something—that they exert some influence on our decisions and behaviors. Fair enough. What we do know from a decade of research is that not only do these intuitions impact people's behavior far more than they realize, but their influence is more pronounced when the rational systems of the mind go off-line. That is, we tend to exhibit more bias when we're rushed, tired, or just not thinking, or when the long-term-oriented systems don't have the time or inclination to fight hard to tip the scale back. Let's look at the evidence.

One of the clearest and most compelling examples of how intuitive biases can sway our behavior comes from work by Alexander

Green, Dana Carney, and their colleagues from Massachusetts General Hospital.[9] The team presented more than sixty white physicians with medical information about several African American patients who were experiencing chest pains. The physicians had to recommend whether or not to treat each patient with clot-busting drugs that would reduce the likelihood of a subsequent cardiac failure. But first the physicians completed an IAT gauging their intuitive views about race, as well as a more explicit questionnaire. What Green and his team found was startling. Even though the symptoms and severity of the heart conditions across black and white patients were the same, in this high-stress and high-fatigue hospital environment, physicians whose IATs indicated stronger intuitive biases against African Americans were significantly less likely to recommend potentially lifesaving treatment for black patients than for white ones. What's more, these same physicians didn't report any prejudiced feelings on their questionnaires; they seemed to have no conscious awareness of their bias. But it nonetheless impacted their actions to a disturbing degree.

Similar studies have found that intuitive biases impact hiring practices as well. The economist Dan-Olof Rooth and his colleagues submitted sets of résumés and applications (for actual jobs) to human resources professionals throughout Sweden. The qualifications on each résumé were identical; the only difference was whether the surname of the applicant was Swedish or Arabic. Previously Rooth had managed to assess intuitive bias in the HR workers by paying a subset of them to take part in a study that, unbeknownst to them, measured their bias using an Arab-vs.-Swede IAT. Here again he found that gut-level bias had a huge impact. The higher the bias against Arabs, the fewer Arab applicants that professional picked out of the hundreds of résumés, even though these applicants had the same qualifications as the Swedes.[10] And once again the level of conscious bias the HR workers reported had no correlation with the decisions they made; like the doctors, they had no idea the bias existed.

Unfortunately, these are but a few of many examples. The list goes on and on. From doctors deciding whether or not to administer lifesaving treatments and HR professionals choosing whom to hire to police officers and soldiers deciding whether or not to pull a trigger, the evidence couldn't be more clear. Prejudices can and do shape our behavior without us even realizing it, but perhaps most troubling, the prejudices themselves can emerge in each of us at the drop of a hat.

Not that there's anything wrong with that . . .

At this point you might be protesting that surely we have more control over our own actions than that. We don't *have* to be slaves to the systems of the grasshopper, do we? There must be a way to keep these biases from rising to the surface and influencing how we act. Why not just try to be careful—to say the right things, make decisions rationally, and make extra sure what we're doing isn't offensive or discriminatory? Put brakes on the grasshopper, so to speak. It sounds like a good idea, but unfortunately, this tactic not only is sometimes ineffective but also tends to backfire.

To understand why, you need look no further than one of the most popular TV shows of all time, *Seinfeld*. In one memorable episode, Jerry and George are trying to convince a reporter who is doing a story on Jerry that they aren't a gay couple. The reporter earlier overheard the two playing a prank in which they pretended to be lovers. But the reporter refuses to believe it was a joke, and now, with their manhood in question, desperation is setting in. "We're not gay!" Jerry protests to the reporter. But as soon as the words fly out of his month, he realizes he doesn't want to be taken for a bigot either, so he blurts out the now-famous line: "Not that there's anything wrong with that." "It's okay if that's who you are," he continues, beginning to stumble. "I have many gay friends." "My father's gay," chimes in George. And on the uncomfortable situation goes as the two put their feet further and further into their mouths trying to convince the reporter that while they aren't gay, they aren't biased against gays either.

The reason this scene is so funny and well remembered is because it perfectly captures the fact that, ironically, trying too hard to seem unbiased can make us look like even more of a bigot. Don't believe us? Say you're trying to point out someone in a crowded train station at rush hour. Let's say this person is black, wearing a blue shirt and brown loafers, and most of the people around happen to be white. You probably don't say, "See him over there, the black guy?" even though it would be the quickest way to identify him. Most people would probably say, "See him over there, the guy in the blue shirt?" even if nine out of ten men in the crowd were wearing blue shirts. The reason? Because many of us think that appearing to simply notice race might make us seem like a racist. So we try to appear completely color-blind, even though we clearly aren't. But this strategy can go very wrong, just as it did for Jerry.

In one of the best demonstrations of this phenomenon, Michael Norton from Harvard Business School and his colleagues conducted two experiments.[11] The first was designed to show that even though most of us claim it, true color blindness is a rare thing when it comes to race. The second was designed to show the counterintuitive results of trying to appear color-blind.

In the first experiment, participants were asked either to sort pictures of faces that varied along a number of characteristics, such as race, gender, hair color, facial expression, age, background color of the photo, and so on, or to guess how long it would take them to sort the faces based on each characteristic. As you might expect, the participants' times revealed that categorizing by race turned out to be one of the fastest ways to sort. However, this fact seemed to come as a bit of a surprise to the participants, the majority of whom, when asked to guess how long it would take them to sort the faces based on the different criteria, reported that race would take them the second-longest time (ranking it only before age). It's important to note that it's not that they thought sorting by color would be hard—they all noted that sorting by hair color and background color would be easy. It was just sorting by skin color that would be difficult. What's most likely the case, of course, is that Norton's participants knew they could easily sort by attending to race; they just didn't want it to appear that way. In short, they wanted to appear color-blind.

So what's the big deal here? Many people would say there's nothing wrong with fudging the truth a little in order to seem like a more open-minded person, right? This brings us to the second experiment. Here Norton and his colleagues had the clever idea of having people play a modified version of the board game Guess Who? Unbeknownst to the participants, of course, each one was purposely paired with either a white or black confederate. After a faked random draw, the participant would learn that she'd be assuming the role of questioner. She would then be given a set of thirty-two photos like those in the first experiment, while the answerer (the confederate) was given a set of six photos: three white faces and three black faces. The goal of the game was simple. On each trial, the questioner had to identify the face that the answerer was looking at using as few yes/no questions as possible.

What happened was right in line with the results of the previous experiment. In an attempt to appear unbiased, white participants were much less likely to ask if the photo was of a white or black individual when the person they were playing with was black. Obviously, asking about the race of the face in the photo was one of the most efficient strategies in playing the game (using physical characteristics as clues to guess the photo is the whole point of the game), but concern about seeming racist was enough to make participants avoid the topic like the plague.

Again, you might be thinking, so what? Well, it turns out that going to such lengths to appear color-blind can actually have some very real and unintended consequences for our social relationships. To see how, let's look at what happened next. Norton had independent third parties view the interactions and make judgments about the participants. When he did, the results were as surprising as they were consistent. The greater the attempt to appear color-blind, the more these individuals were judged as unfriendly and aloof. What this means in practice is that the very people who were trying most to appear unbiased in the presence of a minority group member were exactly the same people who were coming off as disengaged, dismissive, and possibly racist. In other words, they were projecting the exact opposite persona of the one they intended.

Taken together, the upshot of all the work we've discussed here is that whether we like it or not, our minds are built to see the world in terms of alliances—us vs. them—and attempts to counteract the resulting biases at the behest of the ant can sometimes be counterproductive. Yet giving the grasshopper free rein isn't a wise strategy either. Doing so can lead to discrimination of the worst kind, especially when what we know about members of other groups is based on misinformation that's been filtered through a biased lens supplied by the media. But if our minds are prone to stereotyping by extrapolating any information we take in, how can we keep prejudices at bay? It's not easy, but one of the best things we can do is to simply interact with as diverse a group of people as possible. The mind is quick at adjusting its expectations. So the more interactions you have with people not "like you," the better your mind will become at carving out mental shortcuts based on their actual (not supposed) attributes and the more variability in their behaviors you will see, which will strengthen the ant's push to urge you to learn about each person as an individual. Navigating these waters can be treacherous. But as a first step, it's imperative to realize that prejudice can be avoided only when you first recognize that you are not immune to it.

9 / True Colors?

Understanding and managing the spectrum of character

By this point, it's become clear that character—yours, ours, anyone's—is much more flexible than most people would think. Throughout this book we've seen examples of regular people (as well as celebrities and politicians) acting in ways that surprise us—and sometimes acting in ways that surprise even themselves. We've shown you that subtle changes in environment or context can lead any of us to be both saints and sinners. This raises two big questions: Why does the mind work this way? And if what we think of as our character really is so malleable and fickle, can such a thing even be said to exist?

The short answer to the first question is that the system works this way because, quite simply, it's the best evolution has been able to come up with. It works well to optimize our lives, except when it doesn't, but it works more often than not. The long answer, though, requires us to differentiate between what is optimal and what is good. Optimal, at least in the evolutionary sense, means surviving to raise kids, who will carry on your genes. For humans, the optimal choice for how to behave usually lies somewhere between short-term and long-term concerns. Sometimes it's useful to maximize immediate, selfish goals—to cheat for gain, to pretend you have higher status to get something you want, to hit someone before he or she hits you. But acting this way too often will quickly make you shunned, and like it or not, humans need each other for survival. So the mind needs systems that favor both selfless *and* selfish behavior. The trick is figuring out which should take the lead at any given time.

Optimizing your character, then, isn't about being "good" all the time. But you can't be "bad" all the time either and still hope to get by. For example, if you always felt compassion and helped others, you might give away everything you had. But if you were never compassionate, perhaps no one would ever help you when you were in need. Likewise, if you always were a hypocrite, no one would ever trust you, but if you were never a hypocrite, you might not be able to take advantage of a new opportunity that came knocking. The point is, we need flexibility, which is why the mind uses the system it does. If navigating our social world were simple, perhaps we could successfully find our way simply by following a set of maxims or commandments. But it's not simple. Thus, with each new situation, how we should act is computed anew based on the needs and expectations of that specific moment in time. It's all aimed at finding the perfect balance between the two competing sides.

If you've ever studied math or architecture, you're probably familiar with the idea of the "golden mean." If you haven't, the golden mean refers to a ratio that has special properties. It's believed that when the elements of art or architecture are constructed using this ratio, they achieve the perfect balance and thus are most pleasing to the human eye. (Indeed, the golden mean can be found in many of the great masterpieces of modern civilization, from Da Vinci's *Mona Lisa* to Dalí's *The Sacrament of the Last Supper*, from the Parthenon to the Great Pyramids.) But the golden mean also has another interesting albeit less well-known property. It's an irrational number, which means that it changes with each added decimal place; it can never be definitively calculated. Finding the sweet spot for optimizing character is similar. There is a point that works best, but we believe that this spot, just like the golden mean, is always being adjusted. We may get close, but that perfect balance point keeps shifting with each new situation, each new bit of information, each subsequent gambit by our inner ant and our inner grasshopper to take the lead and thereby sway our actions toward its goals.

But don't forget that when we talk about optimal character, we're not equating that with virtue, at least not as it's traditionally defined. After all, virtue can mean many things. For Aristotle, virtue meant optimization in the sense in which we're using it. Virtue, he argued, was to be found in the balance between selfish short-term desires and selfless long-term ones. Vice was to be found at either extreme. But for many others, virtue means "good" in the noble or heavenly sense. The only difficulty here, though, is that what qualifies as "good" often changes across cultures and through time. Although most societies and religions argue that generosity and truthfulness are virtues, some also say that killing can be just, and that men and women should be treated differently. Some religious texts themselves even contain direct contradictions of what virtuous behavior is, leaving it to the current set of priests, rabbis,

We've seen throughout this book that the different processes of the mind really do matter with respect to what you do and how you're perceived. But as it turns out, there is also emerging real-world evidence to suggest that simply being able to recognize the factors that can subtly influence your emotions can produce tremendous social benefits. For example, Marc Brackett and his colleagues at Yale instituted a social and emotional learning program they call the RULER in several elementary school classrooms in which teachers also regularly taught units on developing good character. The RULER approach makes teaching children how to recognize and manage their emotions a central part of the regular academic curriculum, so kids in the RULER classrooms were taught various skills related to knowing when and how it's useful to act on emotional intuitions and when it's not. Kids who didn't get the RULER curriculum learned about developing character in the old way—no skills focusing on different psychological strategies, just the usual "here's how a good person acts" stuff. By year's end, the kids involved in RULER were performing better academically *and* were more socially successful than their non-RULER peers, more often demonstrating work habits and social behaviors that were viewed as desirable, adaptable, and competent.[1]

Similarly, recent findings are beginning to show that being able to recognize biases in our emotional intuitions—and to know when to (or not to) override them—is associated not only with greater life and relationship satisfaction but also with advancement at work and increased leadership potential.[2] All this points to the basic fact that knowledge is power, and that character, like any skill, can be learned, assuming you have the right tools at your disposal. We hope that this book will be one such tool to help get you started.

True colors?

Now let's turn to our second question: does character even exist? Throughout this book, we've been showing you how subtle manipulations in contexts or situations can produce unexpected and wild swings in behavior, driving individuals to act seemingly in ways that are "out of character." Given this fact, you might be tempted to conclude that anyone is capable of anything and "character" simply doesn't exist. That's not exactly right. Character does exist, just not in the way you think. The mistakes we make in classifying a person's character or "true colors" are, in fact, very similar to the errors we make in understanding colors in general. You see, most people perceive colors—red, blue, purple—as defined categories. That is, each color has an essence and clear boundaries. "Purpleness" means something unique, and something very different from "yellowness." What we know from science, of course, is that this isn't the case.

As the frequencies (or wavelengths) of light change, what our eyes see goes from red to green to blue to purple. Our brains perceive these different colors as having unique essences, but in reality they are just variations along a single continuum—the only thing that is changing is the wavelength of the light. So although it may be true that certain yellows are easily identifiable as yellows, it's not always that cut-and-dried. What about citrine, for example? Is it yellow or brown? Can it be both? And isn't brown just a mix of light of other wavelengths anyway? The point is, when you begin to look at color more carefully, it quickly becomes clear that there aren't distinct entities, only spots along a continuum of long to short wavelengths. The boundaries for the labels we use can be quite fuzzy. It's the same with character. Our minds "see" different colors of character—noble, sleazy, trustworthy, unreliable—based on certain actions, but then make the mistake of assigning a person that label unequivocally. So if we define a person as noble and then she does something petty, we assume she's acting out of character. In our minds, noble, just like purple, is a distinct category; it can't bleed over into something else.

But as we've seen, character, like color, varies along a continuum—a continuum not of wavelengths but of our psychological needs flanked by processes embodied in the metaphors of the ant and grasshopper. It is true that, based on differences in temperament, culture, and the types of environs they habitually inhabit, certain people may more frequently seem to occupy one spot along the continuum between long- and short-term desires, and their actions may more frequently tend to correspond to one side of the scale. But as we've shown, where people end up at any one moment often depends on the context. It is certainly the case that each of the competing sides will have its day as situations change. What this means is that so-called swings in character are to be expected; exceptions *are* the rule. There are no firm boundaries for character, only a scale that can shift, and shift quickly, moving us to a new "color" along the spectrum of vice and virtue.

This can be a hard thing to wrap our heads around. When our expectations about someone (or even ourselves) are violated—Tiger Woods' affairs, Lisa Nowak's jealousy-fueled road trip, Farron Hall's act of selfless bravery, and so forth—we often feel we've been fooled. We have. But we've been fooled by the way our brains perceive the world, not by the individual actors. Only once we accept that *all* our minds function along this same continuum and that we

Acknowledgments

Books such as this one are a team effort, and we don't mean just us. Many, many people have played important roles both in conducting the experiments described and in helping us to have the time and opportunity to write about them. First and foremost, we'd like to thank our families, who were probably the only people on the planet happier than we were when we finally finished writing. Dave thanks his wife, Amy, and his daughters, for their insights, moral support, and just plain love (as well as for reminding him that sometimes it's more fun to go to the beach than to stare at a blank computer screen). Carlo thanks his wife, Liz, and daughter, Isabella, for the smiles, distractions, and discussions that made this process a pleasure from start to finish, and the family and friends whose diverse and colorful characters have, perhaps unknowingly, inspired much of the work herein.

Beyond our families, we thank the many brilliant, warm, and wonderful people with whom we've had the good fortune to work. In fact, many of the findings you'll read about in this book would not have been discovered if not for the intelligence and dedication of the young scholars who have been a part of the lab. Each of these individual's contributions deserves to be noted, and so we have included a Guide to the Lab at the end of the volume.

We also owe a debt of gratitude to our editor and champion at Crown, Talia Krohn. Talia was the one who, from the first instant, saw the promise in our work and made it her mission to make sure everyone else did as well. Through several incarnations of the manuscript (and lots of red ink), Talia helped us to tell the story we wanted to tell. We also want to thank our agent, Jim Levine, and all the fine folks at the Levine/Greenberg Literary Agency for guiding us successfully through the sometimes Byzantine-seeming world that is the publishing industry. If not for Jim, we might never have taken a shot at writing a book.

Finally, we want to thank the institutions that have made this work possible. Much of the research described in this book has benefited greatly from the financial assistance provided by the National Science Foundation and the National Institute of Mental Health. We also thank Northeastern University for its continued support of the lab and Dave's fantastic colleagues in the Psychology Department for putting up with his daily complaints about writer's block.

Notes

All of the research we discuss in this book has been published in top-tier scientific journals, which means it's been vetted by many of the most demanding people around—our professional colleagues. But part of good science is being able to look at the data for yourself. So in case you ever wonder whether we considered possibility X or controlled for variable Y, or just how long people helped others or how much they punished them, or exactly how we set up procedure Z, you can find all the nitty-gritty details that we (mercifully) didn't put in this book in the papers and articles referenced below.

1 / Saints and Sinners

1. "S.C. Governor Hears Annual Scouting Report from an Eagle," *Augusta Chronicle*, March 3, 2005.

2. J. Sanford, *Staying True* (New York: Random House, 2010), 27.
3. Ibid., 136.
4. "Governor Sanford's Former Chief of Staff Speaks," WTOC radio interview, June 24, 2009, http://www.wtoc.com/global/story.asp?s=10591454.
5. "Jennifer Sanford's Friends Speak Out," CBS *Early Show* interview, August 24, 2009, http://www.cbsnews.com/stories/2009/08/24/earlyshow/main5261438.shtml.
6. J. Sanford, *Staying True* (New York: Random House, 2010), 167.
7. *Vogue* interview, August 2009, http://abcnews.go.com/Politics/jenny-sanford-tells-vogue-mark-sanfords-affair-shocked/story?id=8348720.
8. P. Rozin and E. Royzman, "Negativity Bias, Negativity Dominance, and Contagion," *Personality and Social Psychology Review* 5 (2001): 296–320.
9. For a great discussion on why populations of true altruists are unstable, see R. H. Frank, *Passions Within Reason: The Strategic Role of the Emotions* (New York: W. W. Norton, 1988).
10. J. T. Cacioppo and B. Patrick, *Loneliness: Human Nature and the Need for Social Connection* (New York: W. W. Norton, 2008).
11. Ran Kivetz and Anat Keinan, "Repenting Hyperopia: An Analysis of Self-Control Regrets," *Journal of Consumer Research* 33 (2006): 273–82.
12. I. Kant, *Groundwork of the Metaphysics of Morals* (Cambridge University Press, 1998).
13. W. Mischel, Y. Shoda, and M. L. Rodriguez, "Delay of Gratification in Children," *Science* 244 (1989): 933–38.
14. For a great discussion of the role of self-control in intertemporal choice, see G. S. Berns, D. Laibson, and G. Loewenstein, "Intertemporal Choice—Toward an Integrative Framework," *Trends in Cognitive Sciences* 11 (2007): 482–88.
15. For a seminal discussion of the benefits of altruism and the role of intuitive emotional responses in causing it, see R. L. Trivers, "The Evolution of Reciprocal Altruism," *Quarterly Review of Biology* 46 (1971): 35–57.
16. We'll be providing many examples of these types of phenomena, but a clear example of the one cited here can be found in P. Valdesolo and D. DeSteno, "Manipulations of Emotional Context Shape Moral Judgment," *Psychological Science* 17 (2006): 476–77.

2 / HYPOCRISY VS. MORALITY

1. Interview on *Meet the Press* (NBC), January 24, 1999.
2. J. Green, "The Bookie of Virtue," *Washington Monthly*, June 2003.
3. P. Valdesolo and D. DeSteno, "The Duality of Virtue: Deconstructing the Moral Hypocrite," *Journal of Experimental Social Psychology* 44 (2008): 1334–38.
4. For a nice demonstration of this fact, see D. T. Gilbert and R. E. Osborne, "Thinking Backward: Some Curable and Incurable Consequences of Cognitive Busyness," *Journal of Personality and Social Psychology* 57 (1989): 940–49.
5. Valdesolo and DeSteno, "The Duality of Virtue."
6. P. Valdesolo and D. DeSteno, "Manipulations of Emotional Context Shape Moral Judgment," *Psychological Science* 17 (2006): 476–77.
7. For great examples of this type of effect, see D. DeSteno, R. E. Petty, D. T. Wegener, and D. D. Rucker, "Beyond Valence in the Perception of Likelihood: The Role of Emotion Specificity," *Journal of Personality and Social Psychology* 78 (2000): 397–416.
8. J. D. Greene, R. B. Sommerville, L. E. Nystrom, J. M. Darley, and J. D. Cohen, "An fMRI Investigation of Emotional Engagement in Moral Judgment," *Science* 293 (2001): 2105–8.
9. S. Schnall, J. Haidt, G. L. Clore, and A. H. Jordan, "Disgust as Embodied Moral Judgment," *Personality and Social Psychology Bulletin* 34 (2008): 1096–1109.
10. C.-B. Zhong and K. A. Liljenquist, "Washing Away Your Sins: Threatened Morality and Physical Cleansing," *Science* 313 (2006): 1451–52.
11. S. Sachdeva, R. Iliev, and D. L. Medin, "Sinning Saints and Saintly Sinners: The Paradox of Moral Self-Regulation," *Psychological Science* 20, 4 (2009): 523–28.

3 / SOUL MATE OR PLAYMATE?

1. Taken from Aristophanes' speech in Plato's *Symposium*.
2. Advice and associated statistics showing effect of pictures reported by Match.com, http://international.match.com/help/faqdetail.aspx?sec=10.

3. R. Anderson and S. Nida, "Effect of Physical Attractiveness on Opposite- and Same-Sex Evaluations," *Journal of Personality* 46 (1978): 401–13; I. H. Frieze, J. E. Olson, and J. Russell, "Attractiveness and Income for Men and Women in Management," *Journal of Applied Social Psychology* 21, 13 (1991): 1039–57; S. G. West and T. J. Brown, "Physical Attractiveness, the Severity of the Emergency and Helping: A Field Experiment and Interpersonal Simulation," *Journal of Experimental Social Psychology* 11 (1991): 531–38; J. E. Stewart II, "Defendant's Attractiveness as a Factor in the Outcome of Trials," *Journal of Applied Social Psychology* 10 (1980): 348–61.
4. J. H. Langlois, J. M. Ritter, L. A. Roggman, and L. S. Vaughn, "Facial Diversity and Infant Preferences for Attractive Faces," *Developmental Psychology* 27 (1991): 79–84.
5. J. H. Langlois, J. M. Ritter, R. J. Casey, and D. B. Sawin, "Infant Attractiveness Predicts Maternal Behaviors and Attitudes," *Developmental Psychology* 31 (1995): 464–72.
6. R. Thornhill and S. Gangestad, "Human Facial Beauty: Averageness, Symmetry and Parasite Resistance," *Human Nature* 4 (1993): 237–69.
7. G. Livshits and E. Kobyiiansky, "Fluctuating Asymmetry as a Possible Measure of Developmental Homeostasis in Humans: A Review," *Human Biology* 63 (1991): 441–66.
8. David Buss, *The Evolution of Desire*, 2nd ed. (New York: Basic Books, 2003).
9. I. S. Penton-Voak and J. Y. Chen, "High Salivary Testosterone Is Linked to Masculine Male Facial Appearance in Humans," *Evolution and Human Behavior* 25, 4 (2004): 229–41.
10. S. W. Gangestad and R. Thornhill, "The Analysis of Fluctuating Asymmetry Redux: The Robustness of Parametric Statistics," *Animal Behaviour* 55, 2 (1998): 497–501.
11. K. M. Durante, N. P. Li, and M. G. Haselton, "Changes in Women's Choice of Dress Across the Ovulatory Cycle: Naturalistic and Laboratory Task-Based Evidence," *Personality and Social Psychology Bulletin* 34, 11 (2008): 1451–60.
12. J. K. Maner, M. T. Gailliot, D. A. Rouby, and S. L. Miller, "Can't Take My Eyes Off You: Attentional Adhesion to Mates and Rivals," *Journal of Personality and Social Psychology* 93, 3 (2007): 389–401.
13. Associated Press, "82 Years of 'I Do': World's Longest Marriage?" June 29, 2005, http://www.msnbc.msn.com/id/8405716.

14. M. McIntyre, S. W. Gangestad, P. B. Gray, J. F. Chapman, T. C. Burnham, M. T. O'Rourke, and R. Thornhill, "Romantic Involvement Often Reduces Men's Testosterone Levels—But Not Always: The Moderating Role of Extrapair Sexual Interest," *Journal of Personality and Social Psychology* 91, 4 (2006): 642–51.
15. M. G. Haselton and S. W. Gangestad, "Conditional Expression of Women's Desires and Men's Mate Guarding Across the Ovulatory Cycle," *Hormones and Behavior* 49 (2006): 509–18.
16. G. C. Gonzaga, D. Keltner, E. A. Londahl, and M. D. Smith, "Love and the Commitment Problem in Romantic Relations and Friendship," *Journal of Personality and Social Psychology* 81, 2 (2001): 247–62.
17. K. Cobb and M. Olson, "NASA to Review Process of Screening Astronauts," *Houston Chronicle*, February 7, 2007.
18. Official Nowak family statement issued on February 6, 2007.
19. D. DeSteno, P. Valdesolo, and M. Y. Bartlett, "Jealousy and the Threatened Self: Getting to the Heart of the Green-Eyed Monster," *Journal of Personality and Social Psychology* 91 (2006): 626–41.
20. M. R. Leary and R. F. Baumeister, *The Nature and Function of Self-Esteem* (San Diego: Academic Press, 2000), 1–62.
21. "Teen Guilty of 1st-Degree Murder in Death of Stefanie Rengel, 14," CBC News, March 20, 2009, http://www.cbc.ca/canada/toronto/story/2009/03/20/murder-rengel-trial.html.
22. DeSteno, Valdesolo, and Bartlett, "Jealousy and the Threatened Self."

4 / FROM PRIDE TO HUBRIS

1. A. Morton, *Tom Cruise: An Unauthorized Biography* (New York: St. Martin's Press, 2008); "Tom Cruise—Biography," TalkTalk, http://www.talktalk.co.uk/entertainment/film/biography/artist/tom-cruise/biography/124.
2. "Ages of Man," *People,* November 16, 1998.
3. "Tom Cruise," WestLord, http://www.westlord.com/tom-cruise-biography.
4. B. Svetkey, "The Crucible," *Entertainment Weekly*, December 20, 1996.
5. A. Bandura, *Self-Efficacy: The Exercise of Control* (New York: Freeman, 1997).
6. In case you were wondering, we had worried that the giving of praise

might make people believe their performance was even better than they might think if they had received the good score alone. However, when we assessed people's judgments of how well they believed they did compared to others, receiving praise didn't improve their judgments of their performance. That is, people who received a score and praise judged their abilities to be the same as those who just received a score. Praise, however, functioned to mark the skill as socially important.

7. L. A. Williams and D. DeSteno, "Pride and Perseverance: The Motivational Role of Pride," *Journal of Personality and Social Psychology* 94 (2008): 1007–17.
8. *Hardball*, MSNBC, May 1, 2003.
9. *Countdown with Keith Olbermann,* MSNBC, May 1, 2003.
10. *Face the Nation,* CBS, May 4, 2003.
11. J. L. Tracy and D. Matsumoto, "The Spontaneous Expression of Pride and Shame: Evidence for Biologically Innate Nonverbal Displays," *Proceedings of the National Academy of Sciences* 105 (2008): 11655–60.
12. A. F. Shariff and J. L. Tracy, "Knowing Who's Boss: Implicit Perceptions of Status from the Nonverbal Expression of Pride," *Emotion* 9 (2009): 631–39.
13. L. A. Williams and D. DeSteno, "Pride: Adaptive Social Emotion or Seventh Sin?" *Psychological Science* 20 (2009): 284–88.
14. P. A. Creed and J. Muller, "Psychological Distress in the Labour Market: Shame or Deprivation," *Australian Journal of Psychology* 58 (2006): 31–39.
15. B. Carey, "When All You Have Left Is Your Pride," *New York Times*, April 7, 2009.
16. R. H. Gramzow and G. Willard, "Exaggerating Current and Past Performance: Motivated Self-Enhancement Versus Reconstructive Memory," *Personality and Social Psychology Bulletin* 32 (2006): 1114–25.
17. R. H. Gramzow, G. Willard, and W. B. Mendes, "Big Tales and Cool Heads: Academic Exaggeration Is Related to Cardiac Vagal Reactivity," *Emotion* 8 (2008): 138–44.

5 / Compassionate or Cruel?

1. "Christmas 1914 and World War One," http://www.historylearningsite.co.uk/christmas_1914_and_world_wa.htm.

2. F. B. M. de Waal, "Putting the Altruism Back into Altruism: The Evolution of Empathy," *Annual Review of Psychology* 59 (2008): 279–300.
3. R. L. Trivers, "The Evolution of Reciprocal Altruism," *Quarterly Review of Biology* 46 (1971): 35–57.
4. G. L. Murphy and D. L. Medin, "The Role of Theories in Conceptual Coherence," *Psychological Review* 92 (1985): 289–316.
5. N. O. Rule and N. Ambady, "The Face of Success: Inferences from Chief Executive Officers' Appearance Predict Company Profits," *Psychological Science*, 19 (2008): 109–111.
6. C. Y. Olivola and A. Todorov, "Elected in 100 Milliseconds: Appearance-Based Trait Inferences and Voting," *Journal of Nonverbal Behavior* 34 (2010): 83–110; C. C. Ballew and A. Todorov, "Predicting Political Elections from Rapid and Unreflective Face Judgments," *Proceedings of the National Academy of Sciences* 104 (2007): 17948–53.
7. U. Mueller and A. Mazur, "Facial Dominance of West Point Cadets as a Predictor of Later Rank," *Social Forces* 74 (1996).
8. J. N. Bailenson, S. Iyengar, N. Yee, and N. Collins, "Facial Similarity Between Voters and Candidates Causes Influence," *Public Opinion Quarterly* 72, 5 (2008): 935–61.
9. Ibid.
10. P. Valdesolo and D. DeSteno, "Compassion, Altruism, and Minimal Group Affiliation" (2007), unpublished manuscript.
11. W. H. McNeill, *Keeping Together in Time: Dance and Drill in Human History* (Cambridge, MA: Harvard University Press, 1995).
12. S. S. Wiltermuth and C. Heath, "Synchrony and Cooperation," *Psychological Science* 20 (2009): 1–5; M. J. Hove and J. L. Risen, "It's All in the Timing: Interpersonal Synchrony Increases Affiliation," *Social Cognition* 27 (2009): 949–61.
13. P. Valdesolo and D. DeSteno, "Synchrony and the Social Tuning of Compassion," *Emotion* (forthcoming).
14. A. Waytz, N. Epley, and J. T. Cacioppo, "Social Cognition Unbound: Psychological Insights into Anthropomorphism and Dehumanization," *Current Directions in Psychological Science* 19 (2010): 58–62.
15. Maurice Bridge, "Bystanders Ignore Plight of Burning Homeless Man," CanWest News Service, December 14, 2005, http://www.freerepublic .com/focus/f-news/1540733/posts.

16. L. T. Harris and S. T. Fiske, "Dehumanizing the Lowest of the Low: Neuro-imaging Responses to Extreme Outgroups," *Psychological Science* 17 (2006): 847–53.
17. C. N. DeWall and R. F. Baumeister, "Alone but Feeling No Pain: Effects of Social Exclusion on Physical Pain Tolerance and Pain Threshold, Affective Forecasting, and Interpersonal Empathy," *Journal of Personality and Social Psychology* 91 (2006): 1–15.

6 / Fairness and Trust

1. Evan Buxbaum, "Storeowner: A Little Compassion Changed Would-Be Robber's Life," CNN, December 3, 2009, http://www.cnn.com/2009/US/12/03/convenience.store.compassion/index.html.
2. Kieran Crowley, "Ex-Thug Repaid Deli Owner Who Helped Him," *New York Post*, December 3, 2009, http://www.nypost.com/p/news/local/he_kept_the_change_3mewgRcqMr311EPvag4sjL.
3. R. H. Frank, *Passions Within Reason: The Strategic Role of the Emotions* (New York: W. W. Norton, 1988).
4. M. Y. Bartlett and D. DeSteno, "Gratitude and Prosocial Behavior: Helping When It Costs You," *Psychological Science* 17 (2006): 319–25.
5. Such helping of others to whom we don't owe anything can foster upstream reciprocity, which is a fancy word akin to "paying it forward." Research has shown that upstream reciprocity can underlie major growth in the levels of cooperation exhibited by members of societies. M. A. Nowak and S. Roch, "Upstream Reciprocity and the Evolution of Gratitude," *Proceedings of the Royal Society B: Biological Sciences* 274 (2007): 605–9.
6. Steve Wulf and Tom Witkowski, "The Glow from a Fire," *Time*, January 8, 1996.
7. David Lamb, "Ethics, Loyalty Are Tightly Woven at Mill," *Los Angeles Times*, December 19, 1996, http://articles.latimes.com/1996-12-19/news/mn-10581_1_malden-mills.
8. D. DeSteno, M. Bartlett, J. Baumann, L. Williams, and L. Dickens, "Gratitude as Moral Sentiment: Emotion-Guided Cooperation in Economic Exchange," *Emotion* 10 (2010): 289–93.

9. S. B. Algoe, J. Haidt, and S. L. Gable, "Beyond Reciprocity: Gratitude and Relationships in Everyday Life," *Emotion* 8 (2008): 425–29.
10. N. M. Lambert, M. Clark, J. Durtschi, F. D. Fincham, and S. Graham, "Benefits of Expressing Gratitude: Expressing Gratitude to a Partner Changes the Expresser's View of the Relationship," *Psychological Science* 21 (2010): 574–80.
11. Kathy Slobogin, "Survey: Many Students Say Cheating's OK," CNN, April 5, 2002, http://archives.cnn.com/2002/fyi/teachers.ednews/04/05/highschool.cheating.
12. F. Gino, S. Ayal, and D. Ariely, "Contagion and Differentiation in Unethical Behavior: The Effect of One Bad Apple on the Barrel," *Psychological Science* 20 (2009): 393–98.
13. C.-B. Zhong, V. B. Lake, and F. Gino, "A Good Lamp Is the Best Police: Darkness Increases Dishonesty and Self-Interested Behavior," *Psychological Science* 21 (2010): 311–14.

7 / Playing It Safe vs. Taking a Gamble

1. Alexandra Berzon, "The Gambler Who Blew $127 Million," *Wall Street Journal,* December 5, 2009, http://online.wsj.com/article/SB125996714714577317.html.
2. A. Tversky and D. Kahneman, "Availability: A Heuristic for Judging Frequency and Probability," *Cognitive Psychology* 5 (1973): 207–32; B. Combs and P. Slovic, "Newspaper Coverage of Causes of Death," *Journalism Quarterly* 56 (1979): 837–43.
3. P. H. Ditto, D. A. Pizarro, E. B. Epstein, J. A. Jacobson, and T. K. MacDonald, "Visceral Influences on Risk Taking Behavior," *Journal of Behavioral Decision Making* 19 (2006): 99–113.
4. S. Gangestad and D. Buss, "Pathogen Prevalence and Human Mate Preferences," *Ethology and Sociobiology* 14 (1993): 89–96.
5. D. Ariely and G. Loewenstein, "The Heat of the Moment: The Effect of Sexual Arousal on Sexual Decision Making," *Journal of Behavioral Decision Making* 19 (2006): 87–98.
6. D. A. DeSteno, R. E. Petty, D. T. Wegener, and D. D. Rucker, "Beyond Valence in the Perception of Likelihood: The Role of

Emotion Specificity," *Journal of Personality and Social Psychology* 78, 3 (2000): 397–416.

7. C. M. Kuhnen and B. Knutson, "The Neural Basis of Financial Risk Taking," *Neuron* 47 (2005): 763–70.
8. B. Dillon, *Tormented Hope: Nine Hypochondriac Lives* (Dublin: Penguin, 2009).
9. K. J. Arrow, "The Theory of Risk Aversion," in *Aspects of the Theory of Risk-Bearing* (Helsinki: Yrjo Jahnssonin Saatio, 1965). Reprinted in *Essays in the Theory of Risk Bearing* (Chicago: Markham Publ. Co., 1971), 90–109.
10. A. A. Baird and J. A. Fugelsang, "The Emergence of Consequential Thought: Evidence from Neuroscience," in *Law and the Brain*, ed. S. Zeki and O. Goodenough (New York: Oxford University Press, 2004), 245–58.
11. A. Bechara, H. Damasio, D. Tranel, and A. R. Damasio, "Deciding Advantageously Before Knowing the Advantageous Strategy," *Science* 275 (1997): 1293–95.
12. R. Thaler and C. Sunstein, *Nudge: Improving Decisions About Health, Wealth and Happiness* (New Haven, CT: Yale University Press, 2009).
13. S. M. Banks, P. Salovey, S. Greener, A. J. Rothman, A. Moyer, J. Beauvais, et al., "The Effects of Message Framing on Mammography Utilization," *Health Psychology* 14 (1995): 178–84.
14. D. T. Gilbert, E. C. Pinel, T. D. Wilson, S. J. Blumberg, and T. Wheatley, "Immune Neglect: A Source of Durability Bias in Affective Forecasting," *Journal of Personality and Social Psychology* 75 (1998): 617–38.

8 / Tolerance vs. Bigotry

1. Transcript taken from video released on Wikileaks. Time refers to minutes and seconds in the video clip. Full video and transcript can be found at http://www.collateralmurder.com.
2. J. Correll, B. Park, C. M. Judd, and B. Wittenbrink, "The Police Officer's Dilemma: Using Ethnicity to Disambiguate Potentially Threatening Individuals," *Journal of Personality and Social Psychology* 83 (2002): 1314–29.
3. Stephen G. Bloom, "Lessons of a Lifetime," *Smithsonian*, September 2005, http://www.smithsonianmag.com/history-archaeology/lesson_lifetime.html.

4. D. DeSteno, N. Dasgupta, M. Y. Bartlett, and A. Cajdric, "Prejudice from Thin Air: The Effect of Emotion on Automatic Intergroup Attitudes," *Psychological Science* 15 (2004): 319–24.
5. "Gibson's Anti-Semitic Tirade—Alleged Cover-up," *TMZ,* July 28, 2006, http://www.tmz.com/2006/07/28/gibsons-anti-semitic-tirade-alleged-cover-up.
6. Ed Pilkington, "Mel Gibson Faces Flak Again After Alleged Racist Rant," *Guardian*, July 2, 2010, http://www.guardian.co.uk/film/2010/jul/02/mel-gibson-racist-rant.
7. Jessica Derschowitz, "Whoopi Goldberg Defends Mel Gibson on 'The View,' " *CBS News,* July 13, 2010, http://www.cbsnews.com/8301-31749_162-20010429-10391698.html.
8. In collaboration with Nilanjana Dasgupta, we've also conducted studies showing that anger can enhance existing biases. For example, when people are made to feel anger, their preexisting automatic biases toward groups such as Arabs become greater. N. Dasgupta, D. DeSteno, L. A. Williams, and M. Hunsinger, "Fanning the Flames: The Influence of Specific Incidental Emotions on Implicit Prejudice," *Emotion* 9 (2009): 585–91.
9. A. R. Green, D. R. Carney, D. J. Pallin, L. H. Ngo, K. L. Raymond, L. Iezzoni, and M. Banaji, "The Presence of Implicit Bias in Physicians and Its Prediction of Thrombolysis Decisions for Black and White Patients," *Journal of General Internal Medicine* 22 (2007): 1231–38.
10. D. Rooth, "Implicit Discrimination in Hiring: Real World Evidence," IZA discussion paper no. 2764, Forschungsinstitut zur Zukunft der Arbeit (Institute for the Study of Labor), Bonn, April 2007, ftp://repec.iza.org/RePEc/Discussionpaper/dp2764.pdf.
11. M. I. Norton, S. R. Sommers, E. P. Apfelbaum, P. Natassia, and D. Ariely, "Color Blindness and Interracial Interaction: Playing the Political Correctness Game," *Psychological Science* 17 (2006): 949–53.

9 / True Colors?

1. M. A. Brackett, S. E. Rivers, M. R. Reyes, and P. Salovey, "Enhancing Academic Performance and Social and Emotional Competence with the

RULER Feeling Words Curriculum," *Learning and Individual Differences* (in press).

2. S. E. Rivers, M. A. Brackett, and P. Salovey, "Measuring Emotional Intelligence as a Mental Ability in Adults and Children," in *The Sage Handbook of Personality Theory and Assessment*, ed. G. J. Boyle, G. Matthews, and D. H. Saklofske, 2:440–60 (Los Angeles: Sage, 2008).

Guide to the Lab

Much of the work that has been described in this book stems from the bright ideas and hard work of our many collaborators. We've made it a point to note each of them by name in the text where we've discussed the experiments of which they were a part, because it is a fact certain that each of their contributions was central to the success of the work. Still, we'd like to single out a few individuals for extra note. Over the past decade, Dave's lab at Northeastern has become a national focal point for the study of how emotions guide social behavior. The result of this fact (and in many ways the reason for it in the first place) is that we've had the extraordinary good fortune to be able to work day in and day out with people who are not only incredibly smart and creative but also some of the warmest and most fun people on the planet. For this we count ourselves blessed. So we think it's only fair to acknowledge present and former members of the lab (in alphabetical order).

Monica Bartlett, who took the lead on much of the work on gratitude, is now an assistant professor at Gonzaga University in Spokane, Washington. Besides being a top-notch scholar on the benefits of positive emotions, Monica is renowned at Gonzaga for her teaching. So if you're ever passing through Spokane, stop in to catch one of her lectures.

Jolie Baumann contributed to some of the work on gratitude we discussed. She is currently a senior graduate student in Dave's lab working in two areas: the effects of emotion on perception and the dynamics of trust. As part of her work on trust, Jolie has teamed with Dave and the MIT Personal Robots Group to give voice to the robot Nexi, which, in collaboration with its creator, Cynthia Breazeal, they are using to study how people decide if they can trust a stranger.

Julia Braverman was Dave's first grad student, who, like all first grad students, helped build the embryonic lab from scratch. After leaving Northeastern with a Ph.D. in psychology, she went on to Harvard to get a master's degree in biomedical informatics. Today she's an instructor at Harvard Medical School doing great work on health communication strategies designed to increase public health.

Paul Condon is a new arrival in Dave's lab. Even though Paul hasn't yet had an opportunity to contribute to any of the work we described, as he's been here only a few months, he has had to put up with Dave saying, "I can't meet now, I'm writing a book." For that alone, he deserves an acknowledgment, but, industrious and creative as he is, he's already conducting cutting-edge experiments on the causes and consequences of human compassion.

Leah Dickens contributed to some of the work on gratitude we discussed. Leah, who will be joining the lab in 2011 as a new grad student, was the lab manager and Dave's research assistant in 2009–10. As such, she also spent many hours on the Internet and in the library hunting down facts and articles for this book. Her intelligence and sheer dedication were amazing assets to this project.

Lisa Williams, who was the guiding force behind much of the work on pride, is now a lecturer at the University of New South Wales in Sydney, Australia. Lisa, who is originally from Hawaii, put in many long hours in the cold Boston winters, so we're happy she's living in Sydney with a view of Coogee Beach from her apartment. Lisa continues to do some of the most groundbreaking work around on the social aspects of pride (and manages to show up in Boston during the winter now and again with a tan).

Index

M

N

O

P

About the Authors

David DeSteno is an associate professor of psychology at Northeastern University, where he is also director of the Social Emotions Lab. He is editor of the American Psychological Association's journal *Emotion* and has served as a visiting associate professor of psychology at Harvard University. His work has been featured in the *New York Times, Wall Street Journal, Washington Post, Boston Globe, Scientific American,* and on ABC News and NPR. He has also guest-blogged for the *New York Times Freakonomics* blog.

Piercarlo Valdesolo is an assistant professor of psychology at Claremont McKenna College. His work has appeared in top journals and major news outlets, including the *New York Times, Washington Post, Boston Globe, LA Times,* and *Newsweek,* and he has been awarded fellowships at Harvard University and Amherst College. He is a contributor to the *Scientific American Mind Matters* blog.